SpringerBriefs in Applied Sciences and Technology

SpringerBriefs present concise summaries of cutting-edge research and practical applications across a wide spectrum of fields. Featuring compact volumes of 50–125 pages, the series covers a range of content from professional to academic.

Typical publications can be:

- A timely report of state-of-the art methods
- An introduction to or a manual for the application of mathematical or computer techniques
- A bridge between new research results, as published in journal articles
- A snapshot of a hot or emerging topic
- An in-depth case study
- A presentation of core concepts that students must understand in order to make independent contributions

SpringerBriefs are characterized by fast, global electronic dissemination, standard publishing contracts, standardized manuscript preparation and formatting guidelines, and expedited production schedules.

On the one hand, **SpringerBriefs in Applied Sciences and Technology** are devoted to the publication of fundamentals and applications within the different classical engineering disciplines as well as in interdisciplinary fields that recently emerged between these areas. On the other hand, as the boundary separating fundamental research and applied technology is more and more dissolving, this series is particularly open to trans-disciplinary topics between fundamental science and engineering.

Indexed by EI-Compendex, SCOPUS and Springerlink.

More information about this series at http://www.springer.com/series/8884

João M. P. Q. Delgado · Joana C. Martinho
Ana Vaz Sá · Ana S. Guimarães
Vitor Abrantes

Thermal Energy Storage with Phase Change Materials

A Literature Review of Applications for Buildings Materials

 Springer

João M. P. Q. Delgado
Faculty of Engineering,
  CONSTRUCT-LFC
University of Porto
Porto, Portugal

Joana C. Martinho
Faculty of Engineering,
  CONSTRUCT-GEQUALTEC
University of Porto
Porto, Portugal

Ana Vaz Sá
Faculty of Engineering,
  CONSTRUCT-GEQUALTEC
University of Porto
Porto, Portugal

Ana S. Guimarães
Faculty of Engineering,
  CONSTRUCT-LFC
University of Porto
Porto, Portugal

Vitor Abrantes
Faculty of Engineering,
  CONSTRUCT-GEQUALTEC
University of Porto
Porto, Portugal

ISSN 2191-530X   ISSN 2191-5318   (electronic)
SpringerBriefs in Applied Sciences and Technology
ISBN 978-3-319-97498-9   ISBN 978-3-319-97499-6   (eBook)
https://doi.org/10.1007/978-3-319-97499-6

Library of Congress Control Number: 2018949869

This Springer imprint is published by the registered company Springer Nature Switzerland AG
The registered company address is: Gewerbestrasse 11, 6330 Cham, Switzerland

Preface

Thermal energy storage with phase change materials (PCMs) offers a high thermal storage density with a moderate temperature variation. Building materials with incorporated phase change materials (PCMs) have been found to reduce significantly indoor temperature fluctuations while maintaining desirable thermal comfort sensation.

This review provides an update on various methods that have been investigated by previous researchers to incorporate PCMs into the building structures. The main objective is to optimize these methods by integrating PCM with surrounding wall (gypsum board and interior plaster products), Trombe walls, ceramic floor tiles, concrete elements (walls and pavements), windows, concrete or brick masonry, underfloor heating, ceilings, thermal insulation and furniture and indoor appliances.

Based on phase change state, PCMs fall into three groups: solid–solid PCMs, solid–liquid PCMs and liquid–gas PCMs. Among them, the solid–liquid PCMs are proper for thermal energy storage. The solid–liquid PCMs include organic PCMs, inorganic PCMs and eutectics.

The process of selecting an appropriate PCM is very complicated but crucial for thermal energy storage. The potential PCM should have a suitable melting temperature, desirable heat of fusion and thermal conductivity specified by the practical application. Thus, the methods of measuring the thermal properties of PCMs are very important.

Suitable PCMs and a right incorporation method with building material and latent heat thermal energy storage (LHTES) can be economically efficient for heating and cooling buildings. However, several problems need to be tackled before LHTES can reliably and practically be applied.

Porto, Portugal

Joana C. Martinho
Ana Vaz Sá
Ana S. Guimarães
Vitor Abrantes
João M. P. Q. Delgado

Contents

Chapter 1
Introduction

A PCM is a substance composed for molecules. The principle of the PCM is simple. As the temperature increases, the material changes phase from solid to liquid. The reaction being endothermic, the PCM absorbs the heat. Similarly, when the temperature decreases, the material changes phase from liquid to solid (see Fig. 1.1). The reaction being exothermic, the PCM desorbs the heat [1].

This kind of material has the capacity of storing and releasing amounts of energy in the form of latent heat; latent heat storage can be achieved through the phase changes. The PCM uses the latent heat of phase change to control temperatures within a specific range.

The energy used to alter the phase of the material, given that the phase change temperature is around the desired comfort room temperature, will lead to a more stable and comfortable indoor climate as well as cut-peak cooling and heating loads [2].

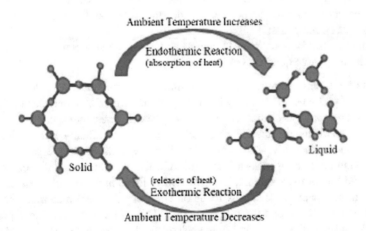

Fig. 1.1 Water melting cycle

© The Author(s), under exclusive license to Springer Nature Switzerland AG 2019
J. M. Delgado et al., *Thermal Energy Storage with Phase Change Materials*,
SpringerBriefs in Applied Sciences and Technology,
https://doi.org/10.1007/978-3-319-97499-6_1

In this chapter, it was presented references of some concepts, definitions and a survey of the different types of PCM.

1.1 Definition

Material properties in general depend on boundary conditions like pressure, temperature and relative humidity. The common materials used in construction, like concrete, brick, stone, glass, wood, ceramic, have a set of properties that give them greater or lesser heat storage capability and heat storage restitution to the surroundings.

Thermophysical properties are those which give information about the amount of energy that such materials and composites can store. But the characterization of the thermophysical properties is not always easy and for composites many times cannot be carried out with conventional laboratory equipment, mostly due to the sample size [3]. The main criteria that oversee the selection of PCMs are [4]:

- Possess a melting point in the desired operating temperature range (temperature range of application) to assure useful heat storage and extraction. Building application temperatures range from 15 °C (cold storage) to 70 °C (heat storage);
- Possess high latent heat of fusion per unit mass, so that a smaller amount of material stores a given amount of energy;
- High specific heat to provide additional significant sensible heat storage effects;
- High thermal conductivity, so that the temperature gradients for charging and discharging the storage material are small;
- Small volume changes during phase transition, so that a simple container and heat exchanger geometry can be used (less than 10%);
- Exhibit little or no sub-cooling during freezing/melting cycle;
- Possess chemical stability, no chemical decomposition and corrosion resistance to construction materials;
- Contain non-poisonous, non-flammable and non-explosive elements/compounds;
- Available in large quantities at low cost.

It is now time to introduce some important concepts related to calorimetry, the part of science that studies energy exchanges in the form of heat between bodies and systems. The concepts of thermal conductivity, specific heat and specific volumetric heat, thermal diffusivity, latent heat or phase change and enthalpy will be addressed.

Thermal conductivity, λ in WmK, describes the transport of energy, in form of heat, through a body of mass as the result of a temperature gradient. According to the second law of thermodynamics, heat always flows in the direction of the lower temperature. For example, let us focus on two very different materials needed to build a wall, concrete and thermal insulation. This material's coexistence provides strength and stability to the building skeleton and at the same time makes it less vulnerable to thermal amplitudes occurring during the day; the amount of heat, per unit time, passes through a thickness unit of material (m), when a temperature unit difference is established between two flat and parallel faces (1 °C or 1 K). Consulting

Table 1.1 Thermal conductivity, density, specific heat, specific volumetric heat of some building materials

Material	Thermal conductivity (λ) [W/(mK)]	Density (ρ) [kg/m^3]	Specific heat (c_p) [kJ/(kgK)]	Specific volumetric heat [kJ/(m^{3}K)]
Water ($T = 10$ °C)	0.600	1000	4.181	4181
Air ($T = 20$ °C)	0.025	1.230	1.012	1.245
Concrete	1.650 2.000	2000–2300 2300–2600	0.880 1.040	1760–2024 2392–2704
Ceramic (roof tiles, bricks, tiles)	0.600	1400–1600	0.840	1176–1344
Mortar (traditional plasters)	1.300 1.800	1800–2000 >2000	1.000 1.046	1800–2000 >2092
Metals (steel)	50.00	7800	0.450–0.512	3510–3994
Wood (dense woods)	0.230	750.0–820.0	1.500–2.500	1125–2050
Calcareous stones	1.400	1800–1990	0.810	1458–1612
Granite	2.800	2500–2700	0.790	1975–2133
Thermal insulation		15.00–20.00		
EPS	0.040	25.00–40.00	1.550	15.83–21.10
XPS	0.037	90.00–140.00	1.045	26.13–41.80
ICB	0.045		0.170	15.30–23.80
Glass (quartz glass)	1.400	2200	0.840	1848

the ITE50 [5], it was found that for concrete $\lambda = 2$ WmK and for thermal insulation $\lambda = 0.04$ WmK, so the difference in conductivity values between these materials indicates that the heat flow through the thermal insulation is fifty times lower than the heat flow that crosses the concrete element for the same superficial area [6].

The specific heat, c_p in units J/kgK, is the amount of heat per unit mass (constant pressure) required to raise the temperature by one degree Celsius. The capacity of the material to storage energy is usually represented by the specific volumetric heat—C, which results from the product of the density by the material's specific heat, $\rho \cdot c_p$, and the units in which it is expressed are in J/m^{3}K [1]. Thermal diffusivity, α_T express in m^2/s, is a material-specific property for characterizing unsteady heat conduction. This value describes how quickly a material reacts to a change in temperature. The metals are materials with a high thermal diffusivity, and they have a fast response to surrounding thermic changes, for example against materials such as expanded polystyrene (XPS), cork and others, which, because of low α_T, can be used as a thermal insulation.

Table 1.1 presents some common materials used in building construction, represented according to their thermic conductivity, density, specific heat and specific volumetric heat [1].

It is important to take into account that although heat storage capacity (that results from the product, $\rho \cdot c_p$) and restitution to stored heat, which is thermic diffusivity, are intrinsic characteristics of the materials, their properties are not constant. In the presence of water, temperature fluctuations or the simply the physical state in which the materials are found can change their properties.

It is well known that all materials interact with the environment; however, most of them lack the capability to alter its own properties according to the environment characteristics in which they are applied. Phase change materials (PCMs), as the name itself advocates, possess the capability to alter its own state as function of the environmental temperature the same does not occur in other traditional building materials [7].

So PCMs are latent heat storage materials, and they use chemical bonds to store and release the heat. The thermal energy transfer occurs when a material changes from solid to liquid, or liquid to solid. This is called a change state or phase. PCMs, having melting temperature between 20 and 32 °C, were used/recommended for thermal storage in conjunction with both active solar storage for heating and cooling in buildings and passive storage, where the heat or cold stored is automatically released when indoor or outdoor temperature rises [8].

Changing of material phase can be classified into four states: solid–solid, solid–liquid, gas–solid and gas–liquid. A phase can be defined as an amount of fully homogeneous material (solid, liquid or gaseous). In each phase can exist at various pressures and temperatures or using the terminology of thermodynamics in several states. The solid phase (a) is characterized by strong molecular cohesion giving stable shape; the liquid phase (b) is an intermediate state; and the gaseous phase (c) by the weak molecular cohesion and therefore without form.

If we consider a given body of water, we recognize that it can exist in various forms. If it is initially liquid may became steam after heating or a solid when cooled. Now consider a given mass of ice, the temperature and pressure known.

According to Coelho [9], when we supply the heat to the ice at constant pressure, the specific volume increases slightly and the temperature rises to 0 °C, at which point ice melts while the temperature remains constant (at 0 °C). At this point, by melting point, all the heat supplied is used to give the change of phase, from solid to liquid. The total amount of heat, or energy, needed to the complete melting fusion of the ice in water is approximately 334 kJ/kg. When all ice is melted, any additional heat transfer causes an increase in the temperature of the liquid (up to 100 °C boiling temperature in the transition from liquid to gaseous). The energy supplied to the water at 0 °C (at the melt temperature) only is restored after the temperature has decreased of water to ice, in an exothermic reaction.

With the illustration of the chart (see Fig. 1.2) below is intended to synthesize the water phases. It represents the energy (in the form of heat) used by water, in the various states, to increase its temperature, allowing to observe that there is a large heat quantity associated with phase changes. In this process, phase change, the amount of energy generated is called "phase shift enthalpy" or only "enthalpy".

The amount of energy required to raise the water temperature from 1 °C to 80 °C is approximately equal to the amount of heat used to melt the ice in water. The amount

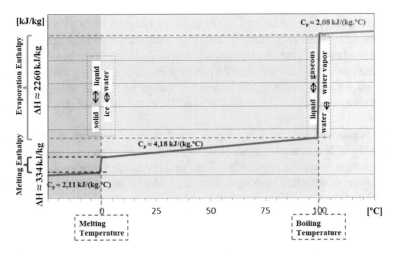

Fig. 1.2 Water phases

of heat associated with the phase change and the amount of heat associated with specific heat c_p of the material are distinguished by latent heat and sensible heat, respectively.

So one of the most important forms of energy storage is thermal energy storage and is applications are very wide, from heating and cooling using waste or solar energy to high-temperature energy storage power production and industrial process. Thermal energy storage (TES) can be stored as a change in internal energy of a material as thermochemical, latent heat and sensible heat or a combination of these. It is important to refer that the concepts of latent heat and sensible heat have a more distinction in this work than the concept thermochemical; a brief reference is described below.

1.1.1 Sensible Heat Storage (SHS)

Sensible heat storage (SHS) method is carried out by adding energy to a material in order to increase its temperature, without changing its phase [10]. The quantity of stored heat depends on the following variables: temperature change, material heat capacity and the amount of storing material. In the SHS method, a solid or a liquid material is used as a storage medium [11], i.e. the medium can be water, bricks, sand, rock beds, oil or soil. Thousands of materials have been identified that are suitable for the use of thermal energy storage. Together with a container an input/output device is attached to it to provide thermal energy for any intended application.

As example, SHS is used as heat storage to provide hot water for houses and offices. Water as storage material has the advantages of being inexpensive and readily available, of having excellent heat transfer characteristics. In solar heating systems, water is still used for heat storage in liquid-based systems, while a rock bed is used for air-based systems [12]. For low- and high-temperature thermal energy storage, solid materials such as metals, rocks, sand, concrete and bricks can also be used [10]. Fernandez et al. [13] have proposed a proper method of selecting the best material to be used for long- and short-term sensible energy storage in order to minimize cost and take into account the availability and environmental aspects such as carbon footprint.

The different storage medium (liquid or solid) presents several advantages and disadvantages. For example, water is widely available and an inexpensive sensible energy storage medium; however, at low- or high-temperature applications, it is limited by the melting and freezing points. To avoid freezing and corrosion problems, chemical additives may be added [14].

1.1.2 Latent Heat Storage (LHS)

Latent heat storage (LHS) is based on the heat release or heat absorption during phase change of a storage material from solid to liquid or liquid to gas or vice versa [15]. There is a visible advantage of PCMs (paraffin wax, salt hydrates and fused salts) over sensible heat storage materials [11, 16]. LHS, compared to SHS, offers higher density of energy storage with near-zero temperature changes. However, difficulties usually arise in real due to the low density change, thermal conductivity, sub-cooling of the phase change materials, stability of properties under extended cycling and sometimes phase segregation [12]. Phase change materials are specifically used in latent heat energy storage systems, and thus, PCM can also be called latent heat storage material. The thermal energy transfer of PCM occurs during the charging or discharging (melting or solidification) process at which the state or phase of the material changes from liquid to solid or from solid to liquid. At the start of the heating of the material, the PCM temperature rises as it absorbs the thermal energy. When the material reaches a specific temperature range, it will start to melt as the material begins to experience a phase transition from solid to liquid state. However, unlike sensible heat storage materials, during the phase transition process, the PCM releases or absorbs heat at a constant or nearly constant temperature. Many authors have experimented with different types of PCMs subdividing them into organic, inorganic and eutectic types. However, the majority of the phase change material does not possess the recommended properties for an ideal thermal energy medium, and thus, thermal enhancers are used to improve any disadvantages that the medium may have. Extensive discussions for each class of phase change material properties can be referred from [15, 17].

1.1.3 Thermochemical

In thermochemical energy storage system, the energy is stored after a breaking or dissociation reaction of chemical bonds at the molecular level which releases energy and then recovered in a reversible chemical reaction. Similar to the other type of thermal energy storage systems, thermochemical heat storage system may also undergo charging, storing and discharging processes. Figure 1.3 illustrates the reversible processes of a thermochemical heat storage system [18]. Additionally, thermochemical heat storage may undergo various processes which include reversible chemical and photochemical reactions, water release from zeolites and hydrates and fuel production. The advantage of this method is that the system is more compact due to the higher energy densities compared to SHS and LHS [11]. Furthermore, the system suffers little or no heat loss during the storing period where the two components, A and B, are stored separately at ambient temperature. Hence, this type of thermal energy storage is more suitable for long-term energy storage, i.e. seasonal storage. In order to select the most suitable candidate for thermochemical heat storage material, several key factors may be used as a rough guideline. These key factors are (i) cost, (ii) ability to sustain large number of charging, storing and discharging cycles, (iii) availability of the material, (iv) non-toxic and non-flammable, (v) corrosiveness, (vi) reaction rate and temperature range, (vii) energy storage density and (viii) good heat transfer characteristics and flow properties.

When using PCMs for TES, the most important material property is the heat storage capability, usually given as the enthalpy as a function of temperature; PCMs have a strong change in enthalpy in a narrow temperature range [19]. In an idealized case, the enthalpy changes suddenly at a phase change temperature. The heat stored is then called latent heat, whereas heat stored with a temperature change is called sensible heat. However, many PCMs change phase in a temperature range, and this

Fig. 1.3 Charging, discharging and storing processes of a thermochemical heat storage system

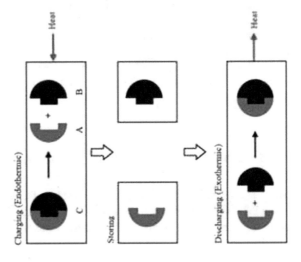

must be taken into account when applying such PCM in a real application. In addition, heating and cooling processes often show different thermal behaviour, called hysteresis. This includes sub-cooling, which means that for the material to change to the lower temperature phase (in a solid–liquid phase change the solid phase), a certain temperature lower than the phase change temperature has to be reached to start the phase change. At this temperature, the nucleation temperature, T_{nuc}, a small nucleus of the lower temperature phase forms. Sub-cooling is very common when using the phase change between solid and liquid [19].

Ice is the best-known PCM used by humans with many and very different types of applications, and for generations, inhabitants of northern Arctic regions have been using ice for thermal stabilization of their dwellings. Igloos are the first-known application of the phase change latent heat in building structures. For example, the igloo is an ingenious invention and very effective in keeping Arctic people warm. Igloos are relatively easy to construct and made from materials found in abundance, snow and ice which serve simultaneously as building structural components, thermal insulation, and thermal radiation shield and energy storage. Blocks of ice are formed into the dome shape, joined together by snow. To prevent excessive amount of snow and cold wind from coming into the igloo, a sunken entrance is constructed, along with a raised sleeping platform covered with fur for comfort and warmth. Internal igloo temperature circulates between 9 and 15 °C, when occupied, even during harsh arctic winters where the outside temperatures can drop to -45 °C [20].

1.2 Classification of PCMs

1.2.1 Non-commercial/Commercial Materials

The selection of an appropriate PCM for any application requires the PCM to have melting temperature within the practical range of application. Several application areas have been proposed for PCMs studied. Table 1.2 presents some companies that commercially produce over 100 PCMs.

In addition to these, several PCMs have been proposed or studied by different researchers. A detailed list of PCMs studied or proposed for study can be found in [21–25].

1.2.2 Organic/Inorganic/Eutectic Materials

There are several materials that can be used as PCMs. A common way to distinguish PCMs is by dividing them into organic, inorganic and eutectic PCMs. These categories are further divided based on the various components of the PCMs (see Fig. 1.4).

Table 1.2 Commercial PCM manufacturers in the world

Manufacturer	PCM temperature range	Number of PCMs listed
RUBITHERM (www.rubitherm.eu)	−3 to 100 °C	29
Cristopia (http://www.cristopia.com)	−33 to 27 °C	12
TEAP (www.teappcm.com)	−50 to 78 °C	22
Doerken (www.doerken.de)	−22 to 28 °C	2
Mitsubishi Chemical (www.m-chemical.co.jp)	9.5 to 118 °C	6
Climator (www.climator.com)	−18 to 70 °C	9
EPS Ltd (www.epsltd.co.uk)	−114 to 164 °C	61

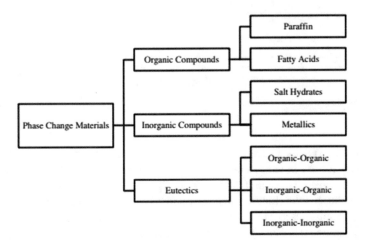

Fig. 1.4 Phase change materials classification

Figure 1.5 shows the difference in melting enthalpy and melting temperature for some of the most common materials used as PCMs.

From the observation of Fig. 1.5, it is possible to conclude that there are three main groups whose melting temperature ranges are compatible with the comfort temperature range inside a building. They are the fatty acids, paraffin, hydrated salts and eutectic mixtures, the latter group having a higher enthalpy range of fusion.

For better understanding and a more complex interpretation of this matter, the subgroups have been described and developed.

1.2.2.1 Organic

Organic phase change materials are divided into paraffin and non-paraffin. In general, organic PCMs do not suffer from phase segregation and crystallize with little or no super-cooling [26].

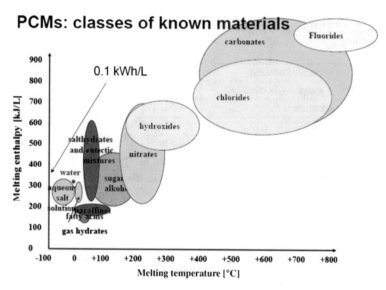

Fig. 1.5 PCM's classes

Paraffin is available in a large temperature range, with a density around 900 kg/m³, opening up for use in various other areas besides building-related applications. The latent heat is mass based, and they show no signs of phase separation after repeated cycling through solid–liquid transitions and have a low vapour pressure (Alkan). However, paraffin used as PCMs has some drawbacks. They have low thermal conductivity (around 0.2 W/mk), and they are not compatible with plastic containers and they are moderately flammable [27].

Non-paraffin used as PCMs includes fatty acids and their fatty acid esters and alcohols, glycols. Fatty acids have received the most attention for use as PCMs in buildings. An extensive review on fatty acids used for PCM purposes has been written by Yuan et al. [28]. Their melting temperatures vary from 5 to 70 °C and possess appreciable latent heat ranging from 45 to 210 J/g but usually around 150 J/g (140 MJ/m³). They have the advantages of congruent melting, low sub-cooling and vapour pressure, non-toxicity, good thermal and chemical stability, small volume change, self-nucleating behaviour and biodegradability. They are also capable of thousands of thermal (melting/freezing) cycles without any notable degradation in thermal properties. Their high surface tension improves their capability of integration in a porous material matrix. However, like paraffin, the major drawback of fatty acids is their low thermal conductivity, ranging from 0.15 to 0.17 W/mk [28, 29].

Esterification of fatty acids with alcohols is a common method to shift the phase transition temperature. It enables decreasing the melting point of fatty acids with high thermal capacity. The production of binary and ternary PCMs by mixing fatty acids with fatty alcohols, polyethylene oxide, oleic acid, pentadecane or other products with low melting temperature is another possible tuning technique [28].

According to Johra [3], other organic PCMs have received less attention by researchers such as sugar alcohol. Some of the polyalcohol has latent heat almost double than that of the other organic PCMs, but their melting point ranges from 90 to 200 °C, which is too high for building applications.

Among them, erythritol is especially noticeable with a latent heat of fusion of 339.8 J/g at 120 °C. Bio-based PCMs are organic materials produced from the biomass: soya bean oils, coconut oils, palm oils and beef tallow. Like the other organic product, they have an interesting latent heat with good chemical stability and phase transition temperatures ranging from −22.77 to 77.83 °C. Nevertheless, they suffer from the same problems as other organic materials [29].

In overall, organic PCMs have many qualities which make them suited for building applications, but there are many organic PCMs considered flammable is a crucial drawback for which impacts the safety aspect of organic PCMs considerably when aimed at building applications.

1.2.2.2 Inorganic

Inorganic phase change materials of interest consist of hydrated salts and metallics. For building applications, however, metallics are not within the desired temperature range and in addition they have severe weight penalties making the unsuited. Hydrated salts consist of an alloy of inorganic salts and water and enable a cost-effective PCM due to easy availability and low cost. The phase change transformation involves hydration or dehydration of the salts in a process that resembles typical melting and freezing. The salt hydrate may either melt to a salt hydrate containing less water or to an anhydrous form where salt and water are completely separated [29].

The salt hydrates possess a significant storage capacity and operate phase transition at ambient temperature. Many studies focused on the calcium chloride hexahydrate, sodium sulphate and magnesium chloride hexahydrate because of their availability, moderate costs and non-flammability. Salt hydrates have a density of around 1700 kg/m^3, which is twice higher than for paraffin. With a maximum latent heat of around 200 J/g, their heat storage on a per volume basis is around 350 MJ/m^3, which is much higher than organic products.

Another significant advantage is their thermal conductivity (around 0.5 W/m K), which is also higher compared to organic materials. However, these products become chemically instable at high temperature. Heating cycles cause continuous dehydration of the PCM, and the heat storage capacity usually degrades over time. Moreover, most salt hydrates melt incongruently with the formation of a lower form product. This irreversible process is an additional drawback for their long-term performance. The liquid phase separation and segregation can be prevented by addition of gelling or thickening agents. Sub-cooling is another problem associated with salt hydrates.

The phenomenon is characterized by a solidification of the product below its phase transition temperature. It can be reduced by inducing heterogeneous nucleation in the salt hydrates, thanks to nucleators or direct contact with an immiscible heat transfer fluid [30].

1.2.2.3   Eutectic Mixtures

A eutectic is a minimum melting composition of two or more components, each of which melts and freezes congruently. During the crystallization phase, a mixture of the components is formed, hence acting as a single component. The components freeze to an intimate mixture of crystals and melt simultaneously without separation [30]. Eutectics can be mixtures of organic and/or inorganic compounds. Hence, eutectics can be made as either organic–organic, inorganic–inorganic or organic–inorganic mixtures [31]. This gives room for a wide variety of combinations that can be tailored for specific applications. Of organic eutectic mixtures, the most commonly tested consist of fatty acids. Some organic eutectics that have been studied include capric acid/myristic acid [32], lauric acid/stearic acid, myristic acid/palmitic acid and palmitic acid/stearic acid [33] and capric acid/lauric acid [34]. The most common inorganic eutectics that have been investigated consist of different salt hydrates. The benefits of eutectic mixtures are their ability to obtain more desired properties such as a specific melting point or a higher heat storage capacity per unit volume. Though it has been given significant interest over the last decade by researchers, the use of eutectic PCMs for use in (LHTS) systems is not as established as pure compound PCMs. Hence, thermophysical properties of eutectics are still a field for further investigations as many combinations have yet to be tested and proved.

In resume, it is possible to see that there is no a perfect product for latent heat thermal energy storage in the temperature range 19–25 °C. Johra [3] shows that very few of them present latent heat above 200 J/g. Organic PCMs offer better chemical and thermal stability with congruent melting, and they exhibit little or no sub-cooling. On the other hand, inorganic products suffer from cycling instability, require nucleating and thickening agents to minimize sub-cooling and are highly reactive to metal materials. Therefore, the organic PCMs (see Table 1.3) seem to be the most appropriate for low-temperature building TES application [3].

To be a desirable material used in latent heat storage systems, the following criteria need to be met: thermodynamic, kinetic, chemical and economic properties, which are shown in Table 1.4.

In conclusion, this chapter presents the concept of PCM and introduces the various types of existing PCMs, divided, in general, into organic, inorganic and eutectic compounds. Organic compounds are seen as the most stable at thermal level.

It has been demonstrated that the principle of operation of PCMs is simple, but that assessing their effectiveness in contributing to latent heat storage and consequently increasing the energy performance of a building can be a challenge.

Table 1.3 Comparison of the different kinds of PCMs

Classification	Advantages	Disadvantages
Organic PCMs	1. Availability in a large temperature range 2. High heat of fusion 3. No super-cooling 4. Chemically stable and recyclable 5. Good compatibility with other materials	1. Low thermal conductivity (around 0.2W/mK) 2. Relative large volume change 3. Flammability
Inorganic PCMs	1. High heat of fusion 2. High thermal conductivity (around 0.5 W/mK) 3. Low volume change 4. Availability in low cost	1. Super-cooling 2. Corrosion
Eutectics	1. Sharp melting temperature 2. High volumetric thermal storage density	Lack of currently available test data of thermophysical properties

Table 1.4 Selection criteria

Thermodynamic properties	(1) Melting temperature in desired range (2) High latent heat of fusion per unit volume (3) High thermal conductivity (4) High specific heat and high density (5) Small volume changes on phase transformation and small vapour pressure at operating temperatures to reduce the containment problems (6) Congruent melting
Kinetic properties	(1) High nucleation rate to avoid super-cooling (2) High rate of crystal growth to meet demands of heat recovery from the storage system
Chemical properties	(1) Complete reversible freezing/melting cycle (2) Chemical stability (3) No degradation after a large number of freezing/melting cycle (4) No corrosiveness (5) No toxic, no flammable and no explosive material
Economic properties	(1) Effective cost (2) Large-scale availabilities

The thermal properties are considered as decisive criterion for choosing the most suitable PCM, highlighting the PCM fusion temperature as the most relevant parameter to make it effective in a given interior environment. Studies argue that the value of the melting temperature should not differ by more than 3 °C relative to the average temperature of a given space.

References

1. A.M. Vaz Sá, Sustentabilidade na construção: comportamento térmico de edifícios em Portugal usando materiais de mudança de fase. Ph.D. Thesis, Faculdade de Engenharia da Universidade do Porto—FEUP, 2013. (in Portuguese)
2. R. Baetens, B.P. Jelle, A. Gustavsen, Phase change materials for building applications: a state-of-the-art review. Energy Build. 1361–1368 (2010)
3. H. Johra, P. Heiselberg, Influence of internal thermal mass on the indoor thermal dynamics and integration of phase change materials in furniture for building energy storage: a review. Renew. Sustain. Energy Rev. 19–32, 2017
4. F. Agyenim, N. Hewitt, P. Eames, M. Smyth, A review of materials, heat transfer and phase change problem formulation for latent heat thermal energy storage systems (LHTESS). Renew. Sustain. Energy Rev. **14**, 615–628 (2010)
5. C. Santos. A.P., L. Matias, Coeficientes de Transmissão Térmica de Elementos da Envolvente dos Edifícios. ICT Informações Científicas e Técnicas, Edifícios - Ite 50, Edited by Laboratório Nacional de Engenharia Civil. LNEC, Lisboa (2007)
6. A. Castilho, Simulação numérica do efeito de PCM no comfort térmico de edifícios – caso de estudo da FEUP. MSc. Thesis, Faculdade de Engenharia da Universidade do Porto - FEUP (2014). (in Portuguese)
7. J. Aguiar, S. Cunha, M. Kheradmand, *Phase Change Materials: Contribute to Sustainable Construction* (2014)
8. V. Tyagi, D. Buddhi, PCM thermal storage in buildings: a state of art. Renew. Sustain. Energy **11**, 1146–1166 (2007)
9. P.J.S.M. Coelho, *Tabelas De Termodinâmica Documentos Técnicos* (FEUP edições, Porto, 2007)
10. M. Telkes, *Trombe Wall with Phase Change Storage Material* (1978)
11. S. Scalat, D. Banu, D. Hawes, J. Paris, F. Haghighata, D. Feldman, Full scale thermal testing of latent heat storage in wallboard. Solar Energy Mater Solar Cells **44**, 49–61 (1996)
12. M. Telkes, *Thermal Energy Storage in Salt Hydrates* (1980), pp. 381–393
13. A.I. Fernandez, M. Martinez, M. Segarra, I. Martorell, L.F. Cabeza, Selection of materials with potential in sensible thermal energy storage. Sol. Energy Mater. Sol. Cells **94**, 1723–9, (2010)
14. B. Farouk, S.I. Guceri, *Tromb–Michal Wall Using a Phase Change Material* (1979)
15. M. Telkes, *Thermal Storage for Solar Heating and Cooling* (1975)
16. C.J. Swet, *Phase Change Storage in Passive Solar Architecture* (1980), pp. 282–286
17. A.A. Ghoneim, S.A. Kllein, J.A. Duffie, *Analysis of Collector—Storage Building Walls Using Phase Change Materials* (1991), pp. 237–242
18. S. Chandra, R. Kumar, S. Kaushik, S. Kaul, *Thermal Performance of a Non-A/C Building with PCM Thermal Storage Wall* (1985), pp. 15–20
19. H. Mehling, L.F. Cabeza, M. Yamaha, *Phase Change Materials: Application Fundamentals. Thermal Energy Storage for Sustainable Energy Consumption* (Springer, Berlin, 2007)
20. J. Kósny, *PCM-Enhanced Building Components: An Application of Phase Change Materials in Building Envelopes and Internal Structures* (Springer, Berlin, 2015)
21. V.V. Tyagi, D. Buddi, *Thermal Cycling Testing of Calcium Chloride Hexahydrate as a possible PCM for Latent Heat Storage* (2008), pp. 891–899
22. G. Lane, *Latent Heat Materials*, vol. 1 (CRC Press, Boca Raton, FL, 1983)
23. M. Kenisarin, K. Mahkamov, Solar energy storage using phase change materials. Renew. Sustain. Energy Rev. **11**, 1913–1965 (2007)
24. P. Verma, Varun, S.K. Singal, Review of mathematical modeling on latent heat thermal energy storage systems using phase-change material. Renew. Sustain. Energy Rev. **12**, 999–1031 (2008)
25. B. Zalba, J.M. Marín, L.F. Cabeza, H. Mehling, Review on thermal energy storage with phase change: materials, heat transfer analysis and applications. Appl. Therm. Eng. **23**(3), 251–283 (2003)

26. M.K. Rathod, J. Banerjee, Thermal stability of phase change materials used in latent heat energy storage systems: a review. Renew. Sustain. Energy Rev. 246–258 (2016)
27. S.D. Sharma, D. Buddhi, R.L. Sawhney, Accelerated thermal cycle test of latent heat storage materials. Sol. Energy **66**, 483–490 (1999)
28. Y. Yuan, N. Zhang, W. Tao, X. Cao, Y. He, Fatty acids as phase change materials: a review. Renew. Sustain. Energy Rev. 482–498 (2014)
29. R.K. Sharma, P. Ganesan, V.V. Tyagi, H.S.C. Metselaar, S.C. Sandaran, Developments in organic solid–liquid phase change materials and their applications in thermal energy storage. Energy Conserv. Manage. 193–228 (2015)
30. L.F. Cabeza, A. Castell, C. Barreneche, A. de Garcia, A.I. Fernández, Materials used as PCM in thermal energy storage in buildings: a review. Renew. Sustain. Energy Rev. 1675–1695 (2011)
31. E. Rodriguez-Ubinas, L. Ruiz-Valero, S. Vega, J. Neila, Applications of phase change material in highly energy-efficient houses. Energy Build **50**, 49–62 (2012)
32. N.S.A. Silva, *Simulação numérica da influência da interface no fenómeno da hu-midade ascensional - Wufi 2D* (Faculdade de Engenharia da Universidade do Porto, Dissertação de Mestrado, 2013)
33. M.N.A. Hawlader, M.S. Uddin, M.M. Khin. *Microencapsulated PCM Thermal-Energy Storage System*, pp. 195–202 (2003)
34. U. Stritih, P. Novak, *Solar Heat Storage Wall for Building Ventilation*, pp. 268–271 (1996)

Chapter 2
Impregnation of PCMs in Building Materials

2.1 Introduction

PCMs utilize the principle of latent heat thermal storage (LHTS) to absorb energy in large quantities when there is a surplus and releasing it when there is a deficit. Correct use of PCMs can reduce peak heating and cooling loads, i.e. reduce energy usage, and may also allow for smaller dimensions of technical equipment for heating and cooling. An added benefit is the ability to maintain a more comfortable indoor environment due to smaller temperature fluctuations [1].

Thermal energy storage with phase change materials (PCMs) offers a high thermal storage density with a moderate temperature variation and has attracted growing attention due to its important role in achievement energy conservation in buildings with thermal comfort [2].

Various methods have been investigated by several researches to incorporate PCMs into building structures, and it has been found that with the help of PCMs the indoor temperature fluctuations can be reduced significantly while maintaining desirable thermal comfort.

Using latent heat storage in the buildings can meet the demand for thermal comfort and energy conservation purpose. This chapter mainly focuses on latent thermal energy storage in building applications, impregnation PCMs into conventional construction materials, current building applications and thermal performance, an introduction to Chap. 3.

2.2 Measurement of Thermal Properties of PCMs

As already mentioned, the process of selecting a suitable PCM is very complicated but crucial for thermal energy storage. The potential PCM should have a suitable melting temperature, desirable heat of fusion and thermal conductivity specified by practical application.

© The Author(s), under exclusive license to Springer Nature Switzerland AG 2019   17
J. M. Delgado et al., *Thermal Energy Storage with Phase Change Materials*,
SpringerBriefs in Applied Sciences and Technology,
https://doi.org/10.1007/978-3-319-97499-6_2

The correct design of the building or storage system with integrated PCMs requires correct knowledge of the thermal properties of the PCMs used. The single data points, the phase change enthalpy at the melting temperature or the heat of fusion do not describe phase change materials properties with accuracy to perform dynamic simulations of a building or a compartment containing PCM [3].

Remembering, the phase change occurs in a temperature range and not at constant temperature level, and therefore, specific heat capacity or enthalpy of this type of material has to be known as a function temperature.

There are many existing measurement techniques, among which differential scanning calorimetry (DSC) and differential thermal analysis (DTA) are most commonly used, nevertheless another measurement technique will be mentioned T-history method.

2.2.1 Differential Scanning Calorimetry (DSC)

One of the most important properties is the enthalpy–temperature relationship, $h(T)$. When this relationship is determined using conventional differential scanning calorimetry (DSC) with standard methods and procedures, results for PCM are often wrong. The enthalpy values from heating/cooling are systematically shifted to higher/lower temperatures. This temperature shift originates from a temperature gradient inside the PCM and depends on the heating/cooling rate and sample mass [4]. There are different possibilities to use a DSC in thermal analysis of PCM, but the most common used are the dynamic method and the step method [5].

The most widely used scanning mode consists of heating and cooling segments at constant rates (dynamic method). A typical temperature program and corresponding signal are shown in Fig. 2.1.

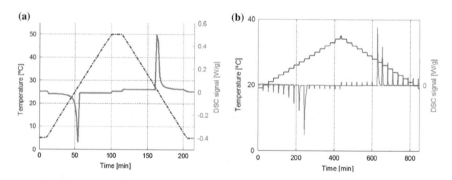

Fig. 2.1 Typical heat flow and temperature evolution during **a** a dynamic DSC measurement with constant heating rate and **b** during DSC measurement with the step method, here shown for heating [5]

So, after a graphic interpretation, the peaks indicate strong thermal effects of the sample at the corresponding temperatures.

According to Cabeza et al. [4], the dynamic method is commonly used for the determination of melting enthalpies. For heat storage applications, the interesting value is the sum of both latent and sensible heat. In this case, good sensitivity also for small signals is necessary. This is achieved using the heat flow rate calibration. For the determination of enthalpy, a dynamic program is executed three times:

- first, with the empty crucible to generate the baseline;
- second, with a standard material (usually sapphire) in the same crucible to generate the standard line;
- third, with the sample in the same crucible to generate the sample line.

From the heat flux, the specific heat as a function of temperature can be obtained with the DSC software using the baseline, the sapphire and the sample heat–flux signal, and the enthalpy is determined by integration.

Pomianowski et al. [3] point out that measurement with different heating rates and that different sample masses gave results that differ considerably from each other and the dynamic mode is not the proper approach can be found and instead of the dynamic mode, an isothermal step mode or T-history method should be used. In the article, it is also mentioned that DSC in general is not suitable for heterogeneous materials, and the shortcomings and some sensitivity analysis of dynamic DSC measurements are discussed and presented in [6].

Another measurement routine is the step method. Here, the heating or cooling is not continuous, but small heating ramps are followed by periods in which the temperature is kept constant to allow the sample to reach thermal equilibrium. The resulting temperature program has small steps, and the signal created is a sequence of different peaks [4].

A typical temperature program and resulting signal is shown in Fig. 2.1. Different peaks indicate different amounts of heat transferred in the respective temperature interval. In the step method, the evaluation considers only peak areas, and the exact shape of the baseline has no influence on the resulting enthalpy–temperature relationship.

The main limitation of DSC is the conditions the sample must fulfil: it should be small, pure and homogeneous. This is a huge limitation because there are many samples that cannot achieve homogeneous conditions since they are a composite materials or a mixture of different components.

Feng et al. [7] summarized the impact of misinterpreted effective capacity function on the simulated thermal storage and releasing effect of the PCM floor. It was found that for the same PCM, the detected results were significantly incongruent. Repeated DSC tests were arranged to discover the influence of heating rate and sample mass on the detected PCM parameters. Errors with 33–883% deviation for phase transition range of PCM were discovered for the improperly arranged tests. These parameters were used in the PCM floor simulation, and a maximum difference of 20% was observed for the floor surface temperature, which greatly influenced the prediction of the simulation. The research shows the importance of setting standard DSC tests and ascertaining right PCM parameters in simulations related to PCM system design.

2.2.2 Differential Thermal Analysis (DTA)

In DTA test, the heat applied to the sample and the reference remains the same (rather than the temperature in DSC test). The phase change and other thermal properties can then be tested through the temperature difference between the sample and the reference.

2.2.3 T-History Method

Zhang and Jiang [8] analysed the limitations of conventional methods including conventional calorimetry, DSC and DTA, and then put forward a new method called T-history method to determine the melting temperature, degree of super-cooling, heat of fusion, specific heat and thermal conductivity of PCMs. They made the measurement of some PCMs through this method and found a desirable agreement between their test results and experimental date available in the literature. Hong et al. [9] modified T-history method by improving some improper assumptions in the method by Zhang and Jiang [8]. Peck et al. [10] also improved this measurement method by setting the test tube horizontally which can minimize the temperature difference along the longitudinal direction of the test tube to get more accurate data from T-history method.

T-history method is based on an air enclosure where temperature is constant and two samples are introduced at a different temperature from the temperature in the air enclosure. During the heating or cooling process, three temperatures are registered, that of the ambient (air enclosure), and those of the two samples. The two samples are one reference substance whose thermal properties are known (frequently water), and one PCM whose thermal properties should be determined with the results of the test.

Marín et al. [11] made improvements in order to obtain enthalpy versus temperature curves. They based their improvements on finite increments method.

Other properties can be studied like sub-cooling and hysteresis analysing the enthalpy–temperature curves. Detailed information about methodology of verification of a T-history installation proposed by University of Zaragoza in collaboration with ZAE Bayern is given by Lázaro et al. [12]. Finally, updates of this methodology can be found in the review published by Solé et al. [13].

2.2.4 Methods Appraisal

The difference between the methods DSC and DTA is that in the first, the energy supplied to the sample and the reference is always the same, with temperature differences between them (sample and reference) occurring whenever a reaction occurs in the sample [14].

The main advantages of the T-history method compared to DSC are [4]:

- precision in energy and temperature measurement;
- sample mass and heating and cooling rate similar to application;
- other properties can be studied like sub-cooling and hysteresis analysing the enthalpy vs. temperature curves.

Although DSC with isothermal step mode and T-history methods are well-developed, they share the same shortcoming—the sample tested has to be homogeneous. In the recent publication [15], a new experimental set-up and various calculation methods to determine the specific heat capacity of inhomogeneous concrete with micro-encapsulated-PCM were suggested. In another publication [16], using experimental data from [15], authors go one step further and compare three different optimization algorithms to define which one is the most effective.

2.3 Thermal Stability of PCMs

For successful large-scale application of PCMs into the building sector, it is crucial that the PCM and PCM-container system can withstand cycling over an extended period of time without showing signs of degradation [1].

There are two main factors which oversee the long-term stability of PCM storage materials: poor stability of the materials, e.g. super-cooling and phase segregation, and corrosion between the PCM and the container system [17, 18].

The long-term stability of the PCMs is required by the practical applications of latent heat storage, and therefore, there should not be major changes in thermal properties of PCMs after undergoing a great number of thermal cycles. Thermal cycling tests to check the stability of PCMs in latent heat storage systems were carried out for organics, salt hydrates and salt hydrate mixtures by many researchers [19–23]. Some potential PCMs were identified to have good stability and thermophysical properties.

Shukla et al. [18] carried out the thermal cycling tests for some organic and inorganic PCMs selected based on thermal, chemical and kinetic criteria, presented on Chap. 1, and their results showed that organic PCMs tend to have better thermal stabilities than inorganic PCMs.

Accelerated ageing tests on stearic acid and paraffin wax, both organic PCMs, have been conducted by Sharma et al. [22]. Both stearic acid and paraffin wax performed well and showed no regular degradation of their melting point over 1500 thermal cycles. However, of the fatty acids, palmitic acid and myristic acid showed to have the best long-term stability [24], which may make them more suited for building applications compared with other fatty acids.

A comprehensive review on the thermal stability of organic, inorganic and eutectic PCMs has recently been given by Rathod and Banerjee [25]. This work covers the investigations on thermal stability of PCMs done over the past few decades. Paraffin has shown good thermal stability. For fatty acids, the purity plays an important role. Industrial grade fatty acids may experience changes in its thermal behaviour over time and should be tested by accelerated ageing. Of inorganic PCMs, salt hydrates are the most widely studied. Most studies have shown that the thermal stability of salt hydrates is poor due to phase separation and super-cooling. However, the thermal stability may be improved to a certain extent by introducing gelled or thickened mixtures and suitable nucleating materials. In general, new building materials, components and structures should be examined by accelerated climate ageing [26]; the PCMs are no exception. Furthermore, a robustness assessment may also be performed [27].

Tyagi and Buddi [28] conducted the thermal cycling test for calcium chloride hexahydrate and found minor changes in the melting temperature and heat of fusion, only about 1–1.5 °C and 4% average variation, respectively, during the 1000 thermal cycles. They recommend the calcium chloride hexahydrate be a promising PCM for applications.

2.4   Heat Transfer Enhancement

Most PCMs suffer from the common problem of low thermal conductivities, being around 0.2 W/m K for paraffin wax and 0.5 W/m K for hydrated salts and eutectics, which prolong the charging and discharging periods. Various techniques have been proposed to enhance the thermal conductivities of the PCMs, such as filling high-conductivity particles into PCMs [6], incorporating porous matrix materials into PCMs [29–36], inserting fibrous materials [35], as well as macro- and micro-encapsulating the PCMs [36, 37].

Bugaje [38] reported that the phase change time is one of the most important design parameters in latent heat storage systems and found adding aluminium additives into paraffin wax can significantly reduce the phase change time in heating and cooling processes. However, this method results in weight increasing and high cost of the system. Metal foams manufactured by sintering method, have many desirable characteristics such as low density, large specific surface area, high specific strength-to-density ratio as well as high thermal conductivity. All these desirable properties offered by metal foams make them to be promising in heat transfer enhancement for PCMs.

Boomsma et al. [39] found using open-cell metal foams in compact heat exchangers generated thermal resistances twice and three times lower than the best commercially available heat exchanger tested. Thermal transport in high porosity open-cell metal foams was experimentally and numerically investigated by [40, 41], in which it is found that the effective thermal conductivity increases rapidly as temperature increases and porosity decreases.

Tian and Zhao [42] conducted a numerical and experimental investigation of heat transfer in PCMs enhanced by metal foams, and their experiment showed a significant increase of heat transfer rate. Their numerical simulations employed two-equation non-thermal equilibrium model to account for coupled heat conduction and natural convection, and a good agreement with experimental data was achieved. They reported that metal foams suppress natural convection while promoting heat conduction significantly, with the overall heat transfer rate still being higher than the pure PCMs.

Py et al. [43] impregnated paraffin wax in a graphite matrix by employing capillary forces, and a high thermal conductivity and stable power output were observed. Fukai et al. [44] found carbon fibres improved the heat exchange rate during the charge and discharge processes even when the volume fractions of carbon fibres were only about 1%. Zhou et al. [2] carried out relevant experiments to compare the effects of metal foams and graphite materials on heat transfer enhancement, and the results indicate that both metal foams and expanded graphite can enhance heat transfer rate in thermal storage system, with metal foams showing a much better performance than expanded graphite.

2.5 Impregnation of PCMs into Construction Materials

Various means of PCM incorporation have been investigated in the literature. Hawes et al. [45] considered three most promising methods of PCM incorporation: direct incorporation or impregnation, immersion and encapsulation.

Zhou et al. [2] refers that the melting and freezing temperatures of PCMs varied slightly when being incorporated in building materials. In addition, PCM can be used in the form of a single laminated board and combined with other envelope components [29, 31].

2.5.1 Direct Incorporation or Impregnation

It is the simplest and the most economical method in which liquid or powdered PCM is directly added to building materials such as gypsum, concrete or plaster during production. No extra equipment is required in this method, but leakage and incompatible with construction materials may be the biggest problems [2].

An example of this method is a laboratory-scale energy storage gypsum wallboard produced by the direct incorporation of 21–22% commercial grade butyl stearate (BS) at the mixing stage of conventional gypsum board production [46].

Figure 2.2 shows Ferreira [47] experiment in which the concrete samples were impregnated in paraffin (RT24).

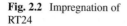

Fig. 2.2 Impregnation of RT24

2.5.1.1 Immersion

In this method, the porous building material (such as gypsum board, brick or concrete block) is dipped into the hot melted PCM, which is absorbed into the pores by capillary action (when a porous material, such as a brick or a wick, is brought into contact with a liquid, it will start absorbing the liquid at a rate which decreases over time, when considering evaporation, liquid penetration will reach a limit dependent on parameters of temperature, relative humidity and permeability) [48].

The porous material is removed from the liquid PCM and allowed to cool, and the PCM remains in the pores of the building material [34]. The great advantage of this method is that it enables one to convert ordinary wallboard to PCM wallboard as required, since impregnation can be carried out at practically any time and place [33].

Hawes and Feldman [35] examined the mechanisms of absorption and established a means of developing and using absorption constants for PCM in concrete to achieve diffusion of the desired amount of PCM. However, as Schossig et al. [36] pointed out, leakage may be a problem over a period of many years for this method.

2.5.2 Encapsulation

To escape the adverse effects of PCMs on the construction material, phase change materials can be encapsulated before incorporation. There are two principal means of encapsulation: macro-encapsulation and micro-encapsulation [17].

2.5.2.1 Macro-encapsulation

The technology with PCMs encapsulated in a container, for example, tubes, spheres or panels, is called macro-encapsulation. The RUBITHERM produces a kind of PCM panels called CSM modules which were made from aluminium with an efficient anti-corrosion coating, shown in Fig. 2.3 [49].

Fig. 2.3 CSM panel
containing the PCM

They can fit many commercial PCMs. With macro-encapsulated PCMs, the leak-age problem can be avoided and the function of the construction structure can be less affected. It has the disadvantages of poor thermal conductivity, tendency of solidification at the edges and complicated integration to the building materials [48].

Via macro-encapsulation, Zhang et al. [50] developed and tested a frame wall that integrated highly crystalline paraffin PCM. Results showed that the wall reduced peak heat fluxes by as much as 38%.

However, macro-encapsulation has the disadvantage of needing protection from destruction and requires much more work to be integrated into the building structure, and is thus expensive.

2.5.2.2 Micro-encapsulation

Nowadays, micro-encapsulated PCMs have been used in thermal energy storage of buildings. Micro-encapsulation is a technology in which PCM particles are enclosed in a thin, sealed and high-molecular-weight polymeric film maintaining the shape and preventing PCM from leakage during the phase change process (see Fig. 2.4). It is much easier and more economic to incorporate the micro-encapsulated PCMs into construction materials [48].

Hawlader et al. [51] conducted thermal analyses and thermal cycle tests on micro-encapsulated paraffin and found that the micro-encapsulated paraffin still kept its geometrical profile and heat capacity after 1000 cycles. Their investigation captures the influence of different parameters on the characteristics and performance of a micro-encapsulated PCM in terms of encapsulation efficiency, and energy storage and release capacity. Results obtained from a DSC show that micro-capsules pre-pared either by coacervation or by the spray-drying methods have a thermal energy

ISE 5.0kV 34.5mm ×2.00k SE(L)   20.0um

Fig. 2.4 Image of PCM micro-capsules in gypsum plaster. The PCM micro-capsules with an average diameter of 8 μm are homogeneously dispersed between the gypsum crystals [36]

storage/release capacity of about 145–240 kJ/kg. Hence, micro-encapsulated paraffin wax shows a good potential as a solar energy storage material.

Some researchers think that the micro-encapsulated PCMs incorporated in the buildings structures may affect the mechanical strength of the structure [52]. Cabeza et al. [53] designed two concrete cubicles with the same shape and size, one with micro-encapsulated PCMs called Mopcon concrete and the other one without PCMs respectively, in order to find the possibility of using micro-encapsulated PCMs in construction materials to achieve sizable energy conservation without significantly decreasing the mechanical strength of the concrete structures at the same time. They found Mopcon concrete reached a compressive strength over 25 MPa and a tensile splitting strength over 6 MPa which had already met the requirements in general structural purpose. However, the applications of micro-encapsulated PCMs still need further investigation in the aspect of safety, such as fire retardation capability etc.

Recently, National Gypsum produced a kind of wallboard panels with Micronal PCM produced by BASF. This kind of panels is called National Gypsum Thermal-CORE Panel, shown in Fig. 2.5. The melting point and latent capacity are 23 °C and 22 BTU/ft^2, respectively.

One of the most widely studied and disclosed constructive solutions are the use of micro-encapsulated PCMs in gypsum plasterboard [14]. This and other constructive solutions with PCM are developed in Chap. 3.

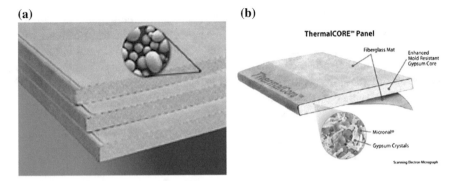

Fig. 2.5 **a** Gypsum wallboard with Micronal PCM (from BASF); **b** thermalCORE phase change drywall (from National Gypsum)

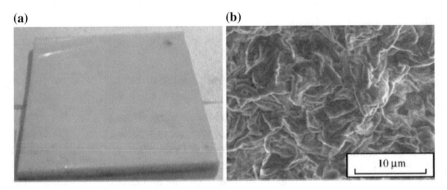

Fig. 2.6 Photos of the shape-stabilized PCM. **a** Photo of the plate and **b** electronic microscopic picture by scanning electric microscope (SEM) HITACHI S-450 [16]

2.5.3 Shape-Stabilized PCMs

In recent years, a kind of novel compound PCM, the so-called shape-stabilized PCM (SSPCM, see Fig. 2.6), has been attracting the interest of researchers [54–60]. It consists of paraffin as dispersed PCM and high-density polyethylene (HDPE) or other material as supporting material.

Since the mass percentage of paraffin can be as much as 80% or so, the total stored energy is comparable with that of traditional PCMs.

So, shape-stabilized PCM is attracting increasing attention due to their large apparent specific heat, suitable thermal conductivity, the ability to keep the shape of PCM stabilized in phase change process, as well as a good performance of multiple thermal cycles over a long period [56–58]. Zhang et al. [54] considered the shape-stabilized PCM and found that it can make the thermal storage system simpler as it does not need special devices or containers to encapsulate the PCM. Based on the above benefits of this shape-stabilized PCM, they also proposed its potential application in efficient buildings used as inner linings, such as inner wall, ceiling and floor. Zhou et al. [55] simulated the thermal performance of a middle direct-gain room with the shape-stabilized PCM plates as inner linings and examined several influencing factors to thermal performance such as melting temperature, heat of fusion, location and board thickness of the shape-stabilized PCM. Their results indicated the PCM plates were advantageous in direct-gain passive solar houses.

2.5.4 Containers

The conventional construction materials, such as gypsum board, concrete, brick and plaster, can be used to hold PCMs. Some other panels, such as PVC panels, CSM panels, plastic and aluminium foils, can also be used to encapsulated PCMs. This subject is developed in Chap. 3.

2.6 Potential PCMs for Building Applications

As we know, many factors influence the indoor air temperature of a building. These include climate conditions (outdoor temperature, wind velocity, solar radiation and others), building structure and the building material´s thermophysical properties (wall thickness, area ratio of window to wall, thermal conductivity and specific heat of wall material), indoor heat source, air change rate per hour and auxiliary heating/cooling installations [48].

Zhang et al. [61] show that the difference between the indoor temperature and the comfort range determines the heating and cooling load when there is no space heating and cooling. Therefore, the heating and cooling load will decrease with decreasing this temperature difference. For a certain building placed in a specific region, the building structure parameters such as wall thickness, area ratio of window to wall, cubage of the room, are known, however the outdoor temperature and solar energy change with the different hour and day during the entire year. Then, with a certain interior heat source, the natural room temperature (i.e. the room temperature without any active cooling or heating) depends on the material thermophysical properties (thermal conductivity, λ, and specific volumetric heat, C (the product of the density by the material's specific heat, $\rho \cdot c_p$).

If there are certain building materials whose λ and $\rho \cdot c_p$ values can make the given room meet the condition $I_{win} = I_{sum} \approx 0$, (being I_{win}, integrated discomfort level for indoor temperature in winter and I_{sum}, integrated discomfort level for indoor temperature in summer), we can call these materials ideal building materials. This means that the indoor temperature will be in the comfort range all year round without auxiliary heating or cooling. [62], also, shows the comparison between the ideal material (in reality, it is very difficult to find this kind of material) and concrete buildings.

The application of PCMs in building can have two different goals. first, using natural heat, that is solar energy, for heating or night cold for cooling; second, using artificial heat or cold sources. In any case, storage of heat or cold is necessary to match availability and demand with respect to time and also with respect power. Basically, there are three different ways to heating or cooling a building. They are:

- PCMs in building walls;
- PCMs in other building components other than walls;
- PCMs in heat and cold storage units.

The first two options are passive systems, where the heat or cold stored is automatically realized when indoor or outdoor temperature rises or falls beyond the melting point. The last one is an active system, where the stored heat or cold is in containment thermally separated from the building by insulation, so the heat or cold is only used on demand not automatically. Some authors classified passive systems applications in the building envelope into two main categories, PCM "integrated" and PCM as "component". The main difference between them is that component can be manufactured before the building being constructed and have a particular design. For example, blinds with integrated PCM are considered as component [63].

Moreover, Kalnæs and Jelle [1] presented many examples of integration of phase change materials for passive systems, exploring possible areas and materials where PCM can be usefully incorporated. Pomianowski et al. [3] presented various construction materials of the building (gypsum and wallboards, concrete, bricks) which where blended or combined with PCM in passive systems. Zhue et al. [64] presented an extensive list of PCM passive systems investigated experimentally with important results. Different possibilities of the use of PCM and their application in the American Solar Decathlon, including the descriptions of the systems and the factors that affect their performance, as well as results of simulations and experimentation were presented by Rodriguez-Ubinas et al. [65]. Soares et al. [66] also explored PCM application in passive systems and investigated the effect of these systems on the energy performance of buildings. Examples of passive system applications are presented and developed in the Chap. 3.

As mentioned above, PCMs incorporated in building envelopes (PCM walls, PCM roof or ceiling and PCM floor) used for passive solar heating in winter can increase thermal capacity of light building envelopes, thus reducing and delaying the peak heat load and reducing room temperature fluctuation. Several PCM applications in buildings such as passive solar heating, active heating and night cooling are shown in Fig. 2.7 [61].

So, PCMs can provide high latent heat thermal energy storage (LHTES) density over the narrow range of temperatures typically encountered in buildings. Therefore, they are taken into account for application.

Thermal comfort can be defined by the operating temperature that varies by the time of the year. The ASHRAE (American Society of Heating, Refrigerating and Air-Conditioning Engineers) has listed suggested temperatures and air flow rates in different types of buildings and environmental circumstances. Normally, the sug-

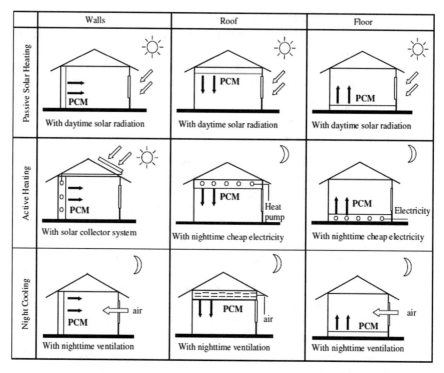

Fig. 2.7 Forms and effects of PCM building envelope [67]

gested room temperature is 23.5–25.5 °C in the summer and 21.0–23.0 °C in the winter. In the building applications, the PCMs with a phase change temperature (18–30 °C) are preferred to meet the need of thermal comfort [2].

As is well known, inorganic PCMs, typically hydrated salts have some attractive properties such as a higher energy storage density, a higher thermal conductivity, being non-flammable, being inexpensive and readily available. However, they also have some obvious disadvantages such as being corrosive, being incompatible with some building materials and needing supporting containers. Some organic PCMs are getting more and more attention due to the avoidance of the problems inherent with inorganic PCMs. They have little super-cooling and segregation, and are compatible with and suitable for absorption in various building materials, though, they are flammable and have volume changes and low heat conductivity, which are concerns in many recent studies [45]. Eutectic or non-eutectic mixtures of organic or inorganic PCMs could be used to deliver the desired melting point required. Shape-stabilized are being applied in building envelopes due to their good thermal performance over a long period, and form stability during heat melting and solidification, which remarkably distinguishes them from common organic PCMs. Also, shape-stabilized PCMs can be easily compounded with common building materials for their shape stability and then can be manufactured into various composite building materials.

In conclusion, with the above methods, extensive advances have been made on the thermal performance of PCM applications in buildings such as PCM walls, PCM ceiling, PCM floor with electric heating and night ventilation.

When selecting PCMs, their phase change temperature should be close to the average room temperature and appropriate values should be required for latent heat and thermal conductivity. Other properties such as fire characteristics and long-term stability should also be considered for organic and inorganic PCMs, respectively. PCMs can be integrated by direct incorporation, immersion, and encapsulation or as a single laminated panel. SSPCM is a promising encapsulation method due to the effectiveness in reducing the danger of leakage as well as its relatively low cost. Thermal analyses showed that PCM walls, floor and ceiling, etc., can be effective in shifting heating and cooling load from peak electricity periods to off-peak periods, or storing solar radiation for use during sunless hours.

References

1. S.E. Kalnæs, B.P. Jelle, Phase change materials and products for building applications: a state-of- the-art review and future research opportunities. Energy Build. **94**, 150–176 (2015)
2. D. Zhou, C.Y. Zhao, Y. Tian, Review on thermal energy storage with phase change materials (PCMs) in building applications. Appl. Energy **92**, 593–605 (2012)
3. M. Pomianowski, P. Heiselber, Y. Zhang, Review of thermal energy storage technologies based on PCM application in buildings. Energy Build. **67**, 56–69 (2013)
4. L.F. Cabeza, C. Barreneche, I. Martorell, L. Miró, S. Sari-Bey, M. Fois, H.O. Paksoy, N. Sahan, R. Weber, M. Constantinescu, E.M. Anghel, M. Malikova, K. Krupa, M. Delgado, P. Dolado, P. Furmanski, M. Jaworski, T. Haussmann, S. Gschwander, A.I. Fernández, Unconventional experimental technologies available for phase change materials (PCM) characterization. Part 1. Thermophysical properties. Renew. Sustain. Energy **43**, 1399–1414 (2015)
5. C. Barreneche, A. Solé, L. Miró, I. Martorell, A.I. Fernández, L.F. Cabeza, Study on differential scanning calorimetry analysis with two operation modes and organic and inorganic phase change material (PCM). Thermochim. Acta **553**, 23–26 (2013)
6. E. Günther, H. Mehling, Enthalpy of phase change materials as a function of temperature: required accuracy and suitable measurement methods. Int. J. Thermophys. **30**, 1257–1269 (2009)
7. G. Feng, K. Huang, H. Xie, H. Li, X. Liu, S. Liu, C. Cao, DSC test error of phase change material (PCM) and its influence on the simulation of the PCM floor. Renew. Energy. 1148–1153 (2016)
8. Y.P. Zhang, Y. Jiang, A simple method, the T-history method, of determining the heat of fusion, specific heat and thermal conductivity of phase-change materials. Measur. Sci. Technol. **10**, 201–205 (1999)
9. H. Hong, S.K. Kim, YS Kim, Accuracy improvement of T-history method for measuring heat of fusion of various materials. Int. J. Refrig. **27**, 360–366 (2004)
10. J.H. Peck, J.J. Kim, C. Kang, H. Hong, A study of accurate latent heat measurement for a PCM with a low melting temperature using T-history method. Int. J. Refrig. **29**:1225–1232 (2006)
11. J.M. Marín, B. Zalba, L.F. Cabeza, H. Mehling, Determination of enthalpy–temperature curves of phase change materials with the temperature-history method: improvement to temperature dependent properties. Meas. Sci. Technol. **14**, 184–189 (2003)
12. A. Lázaro, E. Günther, H. Mehling, S. Hiebler, J.M. Marín, B. Zalba, Verification of a T-history installation to measure enthalpy versus temperature curves of phase change materials. Meas. Sci. Technol. **17**, 2168–2174 (2006)

13. A. Solé, L. Miró, C. Barreneche, I. Martorell, L.F. Cabeza, Review of the T-history method to determine thermophysical properties of phase change materials (PCM). Renew. Sustain. Energy Rev. **26**, 425–436 (2013)
14. A.M. Vaz Sá, Sustentabilidade na construção: comportamento térmico de edifícios em Portugal usando materiais de mudança de fase. Ph.D. Thesis, Faculdade de Engenharia da Universidade do Porto - FEUP, 2013. (in Portuguese)
15. M. Pomianowski, P. Heiselberg, R.L. Jensen, R. Cheng, Y. Zhang, A new experimental method to determine specific heat capacity of inhomogeneous concrete material with incorporated microencapsulated-PCM. Cem. Concr. Res. (2012)
16. R. Cheng, M. Pomianowski, P. Heiselberg, X. Wang, Y. Zhang, A new method to determine thermal physical properties of the mixture of PCM and concrete. Appl. Energy (2012)
17. H. Mehling, L.F. Cabeza, M. Yamaha, *Phase Change Materials: Application Fundamentals. Thermal Energy Storage for Sustainable Energy Consumption* (Springer, Berlin, 2007)
18. A. Shukla, D. Buddhi, R.L. Sawhney, *Thermal Cycling Test of Few Selected Inorganic and Organic Phase Change Materials*. 2606–2614 (2008)
19. S.D. Sharma, D. Buddhi, R.L. Sawhney, Accelerated thermal cycle test of latent heat storage materials. Sol. Energy **66**, 483–490 (1999)
20. K.C. Ting, P.N. Giannakakas, S.G. Gilbert, Durability of latent heat storage tube sheets. Sol. Energy **39**, 79–85 (1987)
21. P.G. Fernanda, Salt hydrate used for latent heat storage: corrosion metals and reliability of thermal performance. Sol. Energy **41**(2), 193–197 (1988)
22. A. Sharma, S.D. Sharma, D. Buddhi, Accelerated thermal cycle test of acetamide, stearic acid and paraffin wax for solar thermal latent heat storage applications. Energy Convers. Manage. **43**, 1923–1930 (2002)
23. H. Kimura, K. Junjiro, Mixture of calcium chloride hexahydrate with salt hydrate or anhydrous salts as latent heat storage materials. Energy Convers. Manage. **28**, 197–200 (1988)
24. A. Sari, K. Kaygusuz, Some fatty acids used for latent heat storage: thermal stability and corrosion of metals with respect to thermal cycling. Renew. Energy. 939–948 (2003)
25. M.K. Rathod, J. Banerjee, Thermal stability of phase change materials used in latent heat energy storage systems: a review. Renew. Sustain. Energy Rev. 246–258 (2016)
26. B.P. Jelle, Accelerated climate ageing of building materials, components and structures in the laboratory. J. Mat. Sci. 6475–6496 (2012)
27. B.P. Jelle, E. Sveipe, E. Wegger, A. Gustavsen, S. Grynning, J.V. Thue, B. Time, K.R. Lisø, J. Build, Robustness classification of materials, assemblies and buildings. Phys. **37**, 213–245 (2014)
28. V.V. Tyagi, D. Buddi, Thermal cycling testing of calcium chloride hexahydrate as a possible PCM for latent heat storage. 2008. 891–899
29. K. Darkwa, J.S. Kim, Heat transfer in neuron composite laminated phase-change drywall. Proc. Inst. Mech. Eng. Part A—J. Power Energy. **218**(A2), 83–88 (2004)
30. J.S. Kim, K. Darkwa, Simulation of an integrated PCM–wallboard system. Int. J. Energy Res. **27**(3), 215–223 (2003)
31. K. Darkwa, J.S. Kim, Dynamics of energy storage in phase change drywall systems. Int. J. Energy Res. **29**(4), 335–343 (2005)
32. I.O. Salyer, A.K. Sircar, *Phase Change Material for Heating and Cooling of Residential Buildings and Other Applications*. Proceedings of the 25th intersociety energy conservation engineering conference (1990), pp. 236–243
33. D. Banu, D. Feldman, F. Haghighat, J. Paris, D. Hawes, Energy-storing wallboard: flammability tests. J. Mater. Civ. Eng. **10**(2), 98–105 (1998)
34. H. Kaasinen, Absorption of phase change substances into commonly used building materials. Sol. Energy Mater. Sol. Cells **27**(2), 173–179 (1992)
35. D.W. Hawes, D. Feldman, Absorption of phase change materials in concrete. Sol. Energy Mater. Sol. Cells **27**(2), 91–101 (1992)
36. P. Schossig, H.M. Henning, S. Gschwander, T. Haussmann, Microencapsulated phase-change materials integrated into construction materials. Sol. Energy Mater. Sol. Cells **89**(2–3), 297–306 (2005)

37. D.A. Neeper, Solar buildings research: what are the best directions? Passive Sol. 213–219 (1986)
38. I.M. Bugaje, Enhancing the thermal response of latent heat storage systems. Int. J. Energy Res. **21**, 759–766 (1997)
39. K. Boomsma, D. Poulikakos, F. Zwick, Metal foams as compact high performance heat exchangers. Mech. Mater. **35**, 1161–1176 (2003)
40. Y. Tian, C.Y. Zhao, *Heat Transfer Analysis for Phase Change Materials (PCMs)*. The 11th International Conference on Energy Storage (Effstock 2009), Stockholm, June 2009
41. C.Y. Zhao, W. Lu, Y. Tian, Heat transfer enhancement for thermal energy storage using metal foams embedded within phase change materials (PCMs). Sol. Energy **84**(8), 1402–1412 (2010)
42. Y. Tian, C.Y. Zhao, *Thermal Analysis in Phase Change Materials (PCMs) Embedded with Metal Foams*. International Heat Transfer Conference-14, Washington, D. C., USA, 8–13 Aug 2010
43. X. Py, R. Olives, S. Mauran, Paraffin/porous graphite-matrix composite as a high and constant power thermal storage material. Int. J. Heat Mass Transf. **44**, 2727–2737 (2001)
44. J. Fukai, Y. Hamada, Y. Morozumi, O. Miyatake, Improvement of thermal characteristics of latent heat thermal energy storage units using carbon-fiber brushes: experiments and modeling. Int. J. Heat Mass Transf. **46**, 4513–4525 (2003)
45. D.W. Hawes, D. Feldman, D. Banu, Latent heat storage in building materials. Energy Build. **20**, 77–86 (1993)
46. D. Feldman, D. Banu, D. Hawes, E. Ghanbari, Obtaining an energy storing building material by direct incorporation of an organic phase change material in gypsum wallboard. Solar Energy Materials **22**, 231–242 (1991)
47. H.S. Ferreira, Os materiais de mudança de fase (pcm) no controlo das humidades ascen-sionais em elementos construtivos. Dissertação de mestrado, Faculdade de engenharia da universidade do porto-FEUP (2016)
48. Y. Zhang, G. Zhou, K. Lin, Q. Zhang, H. Di, Application of latent heat thermal energy storage in buildings: state-of-the-art and outlook. Build. Environ. **42**, 2197–2209 (2007)
49. A. Castell, I. Martorell, M. Medrano, G. Perez, L.F. Cabeza, Experimental study of using PCM in brick constructive solutions for passive cooling. Energy Build. 534–540 (2010)
50. M. Zhang, A.M. Mario, B.K. Jennifer. Development of a thermally enhanced frame wall with phase-change materials for on-peak air conditioning demand reduction and energy savings in residential buildings. Int. J. Energy Res. 795–809 (2005)
51. M.N.A. Hawlader, M.S. Uddin, M.M. Khin. Microencapsulated PCM thermal-energy storage system. Appl. Energy 195–202 (2003)
52. A.M. Khudhair, M.M. Farid, A review on energy conservation in building applications with thermal storage by latent heat using phase change materials. Energy Convers. Manag. **45**(2), 263–275 (2004)
53. L.F. Cabeza, C. Castellon, M. Nogues, M. Medrano, R. Leppers, O. Zubillaga, Use of micro-encapsulated PCM in concrete walls for energy savings. Energy Build. **39**, 113–119 (2007)
54. Y.P. Zhang, K.P. Lin, R. Yang, H.F. Di, Y. Jiang, Preparation, thermal performance and application of shape-stabilized PCM in energy efficient buildings. Energy Build. 1262–1269 (2006)
55. G.B. Zhou, Y.P. Zhang, K.P. Lin, W. Xiao, Thermal analysis of a direct-gain room with shape-stabilized PCM plates. Renew. Energy. 1228–1236 (2008)
56. H. Inaba, P. Tu, Evaluation of thermophysical characteristics on shape stabilized paraffin as a solid-liquid phase change material. Heat Mass Transf. 307–312 (1997)
57. M. Xiao, B. Feng, K. Gong, Preparation and performance of shape stabilizes phase change thermal storage materials with high thermal conductivity. Energy Conserv. Manage. 103–108 (2002)
58. A. Sari, Form-stable paraffin/high density polyethylene composites as a solid–liquid phase change material for thermal energy storage: preparation and thermal properties. Convers. Manage. 2033–2042 (2004)
59. H. Ye, X.S. Ge. Preparation of polyethylene-paraffin compound as a form-stable solid–liquid phase change material. Solar Energy Mat. Solar Cells. 37–44 (2000)

60. M. Xiao, B. Feng, K.C. Gong, Thermal performance of a high conductive shape-stabilized thermal storage material. Solar Energy Mat. Solar Cells. 293–296 (2001)
61. Y.P. Zhang, K.P. Lin, Q.L. Zhang, H.F. Di, Ideal thermophysical properties for free-cooling (or heating) buildings with constant thermal physical property material. Energy Build. **38**, 1164–1170 (2006)
62. A. Castilho, Simulação numérica do efeito de PCM no comfort térmico de edifícios – caso de estudo da FEUP. M.Sc. Thesis, Faculdade de Engenharia da Universidade do Porto - FEUP, 2014. (in Portuguese)
63. J. Aguiar, S. Cunha, M. Kheradmand, *Phase Change Materials: Contribute to Sustainable Construction* (2014)
64. N. Zhu, Z. Ma, S. Wang, Dynamic characteristics and energy performance of buildings using phase change materials: a review. Energy Convers. Manage. **50**, 3169–3181 (2009)
65. E. Rodriguez-Ubinas, L. Ruiz-Valero, S. Vega, J. Neila, Applications of phase change material in highly energy-efficient houses. Energy Build. **50**, 49–62 (2012)
66. N. Soares, J.J. Costa, A.R. Gaspar, P. Santos, Review of passive PCM latent heat thermal energy storage systems towards buildings' energy efficiency. Energy Build. **59**, 82–103 (2013)
67. L.F. Cabeza, A. Castell, C. Barreneche, A. de Garcia, A.I. Fernández, Materials used as PCM in thermal energy storage in buildings: a review. Renew. Sustain. Energy Rev. 1675–1695 (2011)

Chapter 3
PCM Current Applications and Thermal Performance

3.1 Introduction

The use of phase change materials (PCM) in the buildings is a possibility to achieve the reduction of energy dependency as it allows the use of latent heat storage to increase the thermal inertia without significantly increasing the building weight.

It was explained in the previous chapters that PCM-enhanced materials function as lightweight thermal mass components of buildings and contribute to reducing energy use in buildings and to the development of "net-zero-energy" buildings through their ability to reduce energy consumption for space conditioning and peak loads [1].

The use of PCM, to ensure the thermal inertia, in addition to the use of thermal insulation and shading systems, allows the reduction of the winter heat losses and summer heat gains. The use of solar gains, night cooling and off-peak electricity will reduce the evening temperature fluctuations and peak temperatures, increasing comfort conditions inside buildings. These measures will lower both annual energy consumption and the maximum power consumption, saving energy and running costs, for both heating and cooling seasons, both in residential or office buildings and have potential for application in retrofit projects [1–3].

The phase change in the PCM takes place over a small temperature span; thus, large amounts of energy can be stored by small temperature change in the PCM [2]. This means that PCM will not absorb any heat from the air until it has reached the desired temperature range; thus, only excess heat will be stored. PCM can be used to store or extract heat without substantial change in temperature. Hence, it can be used for temperature stabilization in a building. The main advantage of PCM is that, depending on the PCM type, it can store about 3–4 times more heat per volume than sensible heat in solids and liquids at an approximate temperature of 20 °C [4].

Between all phase change materials possible applications in buildings, the most interesting is its incorporation in construction materials altering their materials thermal properties. The PCM may be used for thermal storage of passive solar heating

J. M. Delgado et al., *Thermal Energy Storage with Phase Change Materials*,
SpringerBriefs in Applied Sciences and Technology,
https://doi.org/10.1007/978-3-319-97499-6_3

being integrated in the floor, walls or ceilings, as well as being an integrating part of the most complex energetic system, such as heat pumps and solar panels [5].

When selecting a PCM, the average room temperature should be close to the melting/freezing range of the material. Moreover, the day temperature and solar radiation fluctuations should allow the material phase change. Then many factors influence the choice of the PCM: weather, building structure and thermophysical properties [6].

That's why experiments must be carried out to effectively assess the use of PCM. This is one of many examples given in this chapter, Scalat el al. [7] conducted a full-scale thermal storage tests in a room lined with PCM (Emerest 2326) wallboard and the results show its efficient function as a thermal storage medium, the human comfort can be maintained for longer periods using PCM wallboard, after the heating or cooling system was stopped.

In this chapter, we will summarize current building applications and their performance analyses always with references to different authors. These are the PCM building applications:

- Gypsum board and interior plaster products;
- Ceramic floor tiles;
- Concrete elements (walls and pavements);
- Trombe walls;
- Windows;
- Concrete or brick;
- Underfloor heating;
- Ceilings;
- Thermal insulation materials;
- Furniture and indoor appliances.

3.2 Gypsum Board and Interior Plaster Products

Through the last years, various researchers have studied and developed a vast variety of this type of materials. The main purpose of integrating PCM into lightweight construction materials is to increase their thermal mass. As a result, such products could be used to decrease temperature fluctuations in existing and renovated buildings as well new lightweight buildings.

PCM has been successfully incorporated into wall materials such as gypsum wallboard and concrete to enhance the thermal energy storage capacity of buildings with particular interest in passive solar applications, peak load shifting [5]. It is also the most studied, general and suitable solution for implementing PCM into buildings. Figure 3.1 shows PCM gypsum board.

The wallboards are cheap and widely used in a variety of applications, making them very suitable for PCM encapsulation. However, the principles of latent heat storage can be applied to any appropriate building materials. Kedl and Stovall [8]

Fig. 3.1 PCM-enhanced
gypsum board [11]

and Salyer and Sircar [9] used paraffin wax impregnated wallboard for passive solar application. The immersion process for filling the wallboards with wax was successfully scaled up from small samples to full-size sheets. Processes where by this PCM could be incorporated into plasterboard either by post-manufacturing imbibing of liquid PCM into the pore space of the plasterboard or by addition in the wet stage of plasterboard manufacture were successfully demonstrated [10].

Peippo et al. [12] were one of the first to discuss the use of PCM walls for short-term heat storage in direct-gain passive solar applications. The PCM considered was fatty acid. Approximate formulae were presented for optimum phase change temperature and thickness of the PCM wall. And direct energy savings of 5–20% were expected.

In Feldman et al.'s [13] experiment, a tenfold increase of energy storing capability was obtained by the direct incorporation of 21–22% commercial grade BS at the mixing stage of conventional gypsum board production. Feldman et al. [13–15] carried out extensive research on the use and stability of organic compounds for latent heat storage, including fatty acids (capric, lauric, palmitic and stearic), butyl stearate, dodecanol and polyethylene glycol 600. In addition to the studies of their properties, research was also carried out on materials, which act as PCM absorbers.

Shapiro et al. [16] investigated methods for impregnating gypsum wallboard and other architectural materials with PCM. Different types of PCMs and their characteristics were described. The manufacturing techniques, thermal performance and applications of gypsum wallboard and concrete block, which were impregnated with PCMs. Shapiro [17] showed several PCMs to be suitable for introduction into gypsum wallboard with possible thermal storage applications for the Florida climate. These materials were mixtures of methyl-esters, namely methyl palmitate, methyl stearate and mixtures of short chain fatty acids (capric and lauric acids). Although these materials had relatively high latent heat capacity, the temperature ranges required in achieving the thermal storage did not fall sufficiently within the range of comfort for buildings in hot climates.

Various materials were considered, including different types of concrete and gypsum. The utilization of latent heat storage over a comfortable indoor temperature range in buildings can result in an increase in the thermal storage capacity in the range of 10–130%. The PCM gypsum board was made by soaking conventional gypsum board in liquid butyl stearate, a PCM with phase change range of 16–20.8 °C. The PCM gypsum board contained about 25% by weight proportion of butyl stearate. Its thermal properties were measured with a differential scanning calorimeter (DSC). In another study, investigation of the thermal performance and estimation of the benefits from the application of PCM gypsum board in passive solar buildings in terms of the reduction of room overheating and energy savings were done by Hawes et al. [18].

During the 1980s, several forms of bulk encapsulated PCM were marketed for active and passive solar applications, including direct gain. However, the surface area of most encapsulated commercial PCM products was inadequate to deliver heat to the building passively after the PCM was melted by direct solar radiation. In contrast, the walls and ceilings of building offer large areas for passive heat transfer [10]. Neeper [9] in his study concluded that gypsum wallboard impregnated with PCM could be installed in place of ordinary wallboard during new construction or rehabilitation of a building, thereby adding the regarding thermal storage for passive solar heating as well as creating opportunity for ventilate cooling and time-shifting of mechanical cooling loads. Little or no additional cost would be suffered for installation of PCM wallboard in place of ordinary wallboard. Neeper [19] found that the thermal storage provided by PCM wallboard would be sufficient to enable a large solar heating fraction with direct gain. Neeper [20] examined the thermal dynamics of a gypsum wallboard impregnated by fatty acids and paraffin waxes as PCMs subjected to the diurnal variation of room temperature but not directly illuminated by the sun. He found that the diurnal storage achieved in practice may be limited to the range 300–400 kJ/m^2, even if the wallboard has a greater latent capacity. A wide phase transition range would provide less than optimal storage, but would be consistent with application of the same PCM to either interior partitions or to the envelope of the building. The melting temperatures of these PCMs were adjusted by using mixture of ingredients. He examined three parameters of PCMs wallboards that may influence the energy that can be passively absorbed and released during a daily cycle: (a) the melt temperature of the PCM; (b) the temperature range over which melt occurs; and (c) the latent capacity per unit area of wallboard.

Heim and Clarke [21] conducted numerical simulations for a multi-zone, highly glazed and naturally ventilated passive solar building. PCM-impregnated gypsum plasterboard was used as an internal room lining. The results show that solar energy stored in the PCM–gypsum panels can reduce the heating energy demand by up to 90% at times during the heating season.

Several authors investigated methods for impregnating gypsum and other PCMs [22–30]. Limited analytical studies of PCM wallboard have been conducted, but few general rules pertaining to the thermal dynamics of PCM wallboard are available [10]. It was documented that in gypsum materials can be combined up to 45% by weight of PCM when reinforcing the structure with some additives and up to 60% by weight in wallboard composites.

Voelker et al. [31] have developed the gypsum board with integrated micro-encapsulated PCM and mineral aggregates and have added some admixtures to improve working properties of the board. The incorporated PCM had a melting range between 25 and 28 °C. The sensible and latent heat of the material was measured with differential scanning calorimetry (DSC) with a constant heat and cooling rate of 2 K/min. The thermal conductivity of PCM-modified gypsum was determined with use of a laser flash instrument.

The developed PCM boards were tested in the special lightweight chambers. The two identical test chambers were built next to each other, and in the first one, walls were covered with PCM plaster boards and in the second one with ordinary plaster boards. The thickness of the gypsum board was varied between 1 and 3 cm. The test series were carried out under controlled variable conditions. It was discovered that during warm days, a reduction of the peak temperature of about 3 K in comparison to the room without PCM could be achieved. On the other hand, temperature in the test chamber was allowed to fluctuate from very low to very high temperatures (approximately 14–35 °C). In real building conditions, such high-temperature amplitude would not be acceptable and therefore also utilization of the latent heat of PCM in the gypsum boards would be decreased. Additionally, authors do not elaborate on obtained PCM to gypsum ratio in the developed gypsum boards.

Kuznik et al. [32] investigated a renovation project in the south of Lyon, France using PCM wallboards. By testing a room in the same building that was renovated without PCM and then comparing it to the room with PCM, they concluded that the PCM increased the indoor thermal comfort, but it appeared unable to use its latent heat storage capacity for a number of durations due to the incomplete discharge overnight.

Kuznik et al. [33], experimental tests on composite PCM product ENERGAIN® from Dupont de Nemours Society that contained 60% of micro-encapsulated paraffin and rest was copolymer can be found. The thermal conductivity has been measured using a guarded hotplate apparatus, and in the liquid state, it was at 0.22 W/mK and in the solid at 0.18 W/mK. The enthalpy of the composite PCM has been measured using DSC method at heating–cooling rate of 0.05 K/min. Melting peak temperature was obtained at 22.2 °C and freezing peak at 17.8 °C. Composite boards were tested in the specially designed full-scale test room MINIBAT. The test room was equipped with thermal guard surrounding room where the temperature was stabilized at certain temperature of 20.5 °C. The test cell was also equipped with a solar simulator located in the climatic chamber that was attached to the test chamber to simulate external thermal condition. The climatic chamber was separated from the test room with a glass. The temperature inside the climatic chamber could vary between −10 and 40 °C and could be dynamically controlled so that any temperature evolution can be generated. In the investigation, three types of test were conducted:

- A summer day: temperature in the climatic chamber varied between 15 and 30 °C, there was night cooling to improve PCM storage/release effect;
- A mid-season day: temperature in the climatic chamber varied between 10 and 18 °C;

- A winter day: temperature in the climatic chamber varied between 5 and 15 °C, heating system in the test room was turned on when temperature in the room dropped below 20 °C.

For all tested cases, the solar flux is preserved as the same. The experiment was conducted in a comparative manner for the test room with PCM and without PCM boards on the walls. Based on the results obtained, authors concluded that PCM composite is an interesting solution for the building application to enhance the human thermal comfort due to three reasons:

- The PCM included in the walls reduced the overheating effect, and energy stored was released to the room when temperature was minimum;
- The wall surface temperature peaks were flattened;
- The stratification of air temperature in the room with PCM was not observed as it was for room without PCM.

Still the allowed temperature fluctuations for the tests were very high. For example, for a summer day the air temperature was allowed to fluctuate between approximately 19 and 32 °C for the room without PCM and from 19 to 29 °C for the room with PCM. Based on that, it can be concluded that if the test rooms were equipped with some additional measures to reduce temperature fluctuations, for example solar shadings, the utilization of PCM in the room would be smaller and also the improvement with regards to the room without PCM would drop.

Athienitis et al. [34] performed an experimental and numerical simulation study in a full-scale outdoor test room with PCM gypsum board as inside wall lining. The PCM gypsum board used contained about 25% by weight proportion of BS. An explicit finite-difference model was developed to simulate the transient heat transfer process in the walls. It was shown that utilization of the PCM gypsum board may reduce the maximum room temperature by about 4 °C during the day and can reduce the heating load at night significantly.

In Schossig et al. [35], measurements of a full-size room equipped with micro-encapsulated-PCM plaster boards are presented. Prior to the full-scale measurements, some small-scale experiments with specially designed plate apparatus to test wall samples have been conducted. A small sample of 50×50 cm^2 area was pressed between two copper plates, which can be heated and cooled independently. The thermal performance of the wall samples with PCM was tested for the constant heat flux on both sides of the sample, and temperature in the middle of the sample was registered. It was discovered that for the samples with PCM temperature instead of rising linearly begins to deflect within PCM melting temperature range. Consecutively, the full-scale measurements have been conducted in the specially built lightweight test rooms. One room was equipped with the ordinary reference plaster and the other with the PCM plaster. Both rooms were facing south. In the article, it was not written how much of internal area was covered with gypsum and where the plaster boards were located. Within the project, two different PCM products were tested: dispersion-based plaster with 40% weight PCM and 6 mm thickness and gypsum plaster with 20% weight PCM and 15 mm thickness. The experimental study indicated that PCM gypsum helped to decrease high- and low-temperature peaks. Over period of 3 weeks, the reference room was warmer than 28 °C for about 50 h while the PCM room was only 5 h above 28 °C. Authors pointed out that micro-encapsulated PCM has the advantage of easy application and there is no danger of leakage like with macro-encapsulated PCM.

Gypsum products as construction materials can be improved by improving their physical, mechanical, thermal, and sound insulation properties. Jeong et al. [36] conducted a study manufactured heat storage gypsum board which contains two types of SSPCMs. Each type of SSPCM has different phase change temperatures for targeting the heating and cooling seasons. Paraffinic organic PCM-based SSPCM and fatty acid-based SSPCM were prepared by using exfoliated graphite nanoplatelets (xGnP) to solve the leakage problem, retaining their efficient thermal storage quantity and improving the thermal conductivity. The two types of SSPCM composites (Hybrid SSPCM) were incorporated to make heat storage gypsum board. The hybrid SSPCM was manufactured for reducing heating and cooling load in severe season such as winter and summer by two types of PCMs with different phase change temperature.

In this experiment, the maximum mix ratio of SSPCM was considered as 30 wt% in the light of workability and banding strength of heat storage gypsum board. All samples were made with a composition of gypsum and 10, 20 and 30 wt% of hybrid SSPCM, in comparison with the gypsum powder weight. For the preparation of hybrid SSPCM gypsum board, the water ratio at 45% was selected, in comparison with gypsum powder. The prepared hybrid SSPCM has a board shape of 100×100 mm $\times 20$ mm^3 (length, width and height).

Bharat Chhugani et al. [37] investigate the effectiveness of PCM wallboards in lightweight buildings. The investigations have been carried out on two different types of PCM wallboards, the Knauf Comfortboard-23 and DuPont Energain board. Experimental results showed that PCM wallboards can provide passive cooling powers of around 8 W/m2 under typical office room conditions. The experiments proved that PCM wallboards can almost store twice as much heat compared to conventional gypsum boards and can provide a passive cooling power which is comparable to a concrete wall with a thickness of 15 cm. However, the regeneration behaviour of PCM wallboards plays a major role in its efficiency. The results reveal that Knauf Comfortboard-23 shows a better regeneration behaviour than the DuPont Energain board. Still, the average regeneration rate of the Comfortboards-23 during the summer months in the offices of the Energy Efficiency Centre was found to be below 20%. The regeneration of the DuPont Energain board was nearly impossible being 1% in average.

To conclude, the realistic potential to increase dynamic heat storage capacity of concretes by incorporation of PCM is doubtful. Firstly, the thermal mass increase is not as high as expected and secondly, thermal conductivity decreases significantly due to addition of PCM to concrete. As a result, the energy from the air has difficulty in being transported to the inside of PCM concrete construction within daily realistic indoor temperature variations. Moreover, maximum amount of PCM in the concrete is not higher than 5–6% by weight (material is still workable), which means not much latent heat capacity can be introduced to sensible heat storage capacity.

Consequently, 5–6% by weight of PCM corresponds to approximately 12–15% by volume of concrete, which means that the share of PCM in concrete is rather high and as a result, the price of the composite would be high due to rather high price of PCM.

3.3   Ceramic Floor Tiles

Ceramic tiles are an extensively used material for paving, yet the incorporation of PCMs into ceramic tiles has been rather neglected, as observed by Pomianowski et al. [38], in their excellent review.

Cerón et al. [39] reported the production, development and experimental method, to test its performance, of a prototype tile. Nonetheless, the design of the tile was rather complex requiring numerous layers (top stoneware tile, metal sheet, a metal container with PCM and a bottom thermal insulation layer). The developed tiles were tested over a period of 60 days in one side of the solar house placed in Madrid that had door-window towards south. It was observed that high effectiveness was achieved for tiles close to the door-window where the direct solar radiation hits the tiles. The contribution of the tiles in the deeper location, further from the window, was very small. Therefore, the key conclusion was that the scheme should be limited to the portion of the floor that can receive the direct solar radiation. It has been considered essential to make a new prototype of tile that could be placed directly on the framework (see Fig. 3.2) and not on a technical floor, since this is a more usual and cheaper solution for housing. It was also considered necessary to develop the system in such a way that it met the acoustic requirements demanded by the Technical Building Code in Spain (CTE) [39].

The results showed that the shadows on the tile with PCM significantly alter the thermal behaviour of the pavement, by reducing their efficiency in solar energy storage. In order to get a high efficiency in the process of thermal storage, it is important to avoid obstacles that cause significant shadows on the pavement that

Fig. 3.2 Prototype [39]

contains PCM. The shadows caused by the aluminium frame of the window result in a decrease of the surface temperature of the tile down to 6 °C over a period of 1 h, with solar radiation 1000 W/m^2, and this reduces the amount of energy accumulated in that period [39].

Hittle et al. [40] wrote an assignment which proposal consists in substitute micro-encapsulated phase change material for much, if not all of the quartz power to make it. He also was able to replace some of the chips without degrading the tile appearance. The tile with PCM it was named here in after. Putting phase change material in the floor tile dramatically increases its ability to store thermal energy. Also, the energy is stored at a nearly constant temperature. One application is the use of the tile in sunroom floors where it can absorb solar energy during the day and release it at night to reduce mechanical heating. Because agglomerate floor tiles have exceptional wear resistance properties, they are often marketed to institutional clients that have high traffic areas. He came to the conclusion that these tiles have structural properties that are not quite as good as agglomerate tiles without PCMs, but are significantly better than fired clay tiles often used in residential applications. Each prototype tile contains quartz chips and powder, polyester resin and encapsulated paraffin wax. Tiles were made with varying amounts of quartz powder, encapsulated wax and resin. The amount of quartz chips will be held constant to ensure structural integrity of the tile. Resin will be varied from 5 to 10% of the total mixture. Proportions of encapsulated wax and quartz powder will be directly dependent on each other. The ideal case would be a tile containing all phase change material and no quartz powder.

One of the standard agglomerate floor tiles manufactured today consists of quart chips, quartz powder (filler), dyes and a polyester binder. Components are mixed in a giant cement mixer. The mix, appearing and feeling much like damp sand, is placed in a vibrating vacuum assembly to remove all air and to compress the material. Next, the portions of material are heated to cause catalysis. Later, pieces are cut and polished to produce the desired tiles.

In a very intricate paper, the production of novel PCM-ceramic tiles for indoor temperature control was described and developed by Novais et al. [41]. The PCM was directly incorporated into lightweight porcelain stoneware ceramic tiles, without the need for other materials or containers. The incorporation of PCM in ceramic tiles increases buildings' thermal inertia and reduces the indoor space temperature variation by up to 22%. The tiles combine a dense top-layer with a porous bottom-layer. The novelty of this investigation is the development of ceramic tiles with a PCM that was directly included on the porous layer of the lightweight porcelain stoneware tiles. Wood wastes (sawdust) were used as a pore forming agent, which is an environmental friendly approach. The density of these novel ceramic tiles (below 2 g/cm^3) allows their use not only in the floor, but also as wall coverings, which increases the surface area inside the buildings and, therefore, the energy savings.

Two major factors control the PCM-ceramic tiles' thermal performance: (i) PCM load and (ii) PCM-ceramic tile thermal conductivity. Higher PCM content induces narrower temperature fluctuations, yet reduces the PCM-ceramic tile thermal conductivity. The incorporation of 5.4 wt% PCM was found to be the optimal content, which is a rather small amount in comparison with other PCM building materials.

The results confirm the enhancement of thermal insulation with the PCM-ceramic tiles when using a 10-min dwell: 0.9 °C reduction on the minimum temperature and differences exceeding 1 °C on the maximum temperature. The overall indoor space temperature amplitude decreased about 2 °C. However, the minimum temperature reached inside the test cell (~20.5 °C) is insufficient to promote the complete solidification of the PCM, which affects its functionality as pointed out by Voelker et al. [31]. In fact, when the PCM has sufficient time for the occurrence of complete melting and solidification, its performance is improved. For example, with 90-min dwell time, the temperature amplitude in the test cell coated with PCM-ceramic tiles decreased by 2.7 °C. The enhancement of the PCM performance is ensured when its complete melting and solidification [41].

In conclusion, PCM-ceramic tiles present remarkable potential for improving the thermal comfort inside buildings due to their ability to reduce indoor space temperature fluctuations. Additionally, if associated with underfloor electric systems the energy consumption can be transferred for off-peak period providing substantial energy savings.

3.4   Concrete Elements (Walls and Pavements)

The thermal response of concrete walls containing PCMs has been reviewed extensively, and Ling et al. [42] show that among the PCM types, organic PCM and particularly paraffin wax PCM seem to be one of the most suitable latent heat storage materials that can be used in concrete. The main reasons are the chemical stability, inactivity in the alkaline environment of concrete, an appropriate transition temperature of about 26 °C (human thermal comfort) and low degree of super-cooling; they are also relatively inexpensive and have desirable thermal stability. The test results of different means of PCM incorporation in concrete showed that: Immersion: suitable for concrete with a relatively high porosity. The time required for immersion is mainly controlled by (i) the absorption capacity of the porous concrete and (ii) the temperature of the container in which the melted (liquid) PCM is filled. Basically, the immersion process takes several hours. Impregnation: vacuum impregnation seems to be more effective compared to the simple immersion technique. By comparing the results of absorption behaviour of PCM in different types of porous aggregates, expanded clay or shale aggregates are the more suitable porous materials for PCM impregnation. Direct mixing: encapsulation of the PCM with a chemically and physically stable shell is required before it can be directly mixed into concrete. The surface (shell) hardness of the PCM micro-capsules can be reinforced by the use of zeolite or zeocarbon.

In brief, most experimental studies of walls or rooms exposed to outdoor conditions have reported that adding PCM to building walls reduced the amplitude of the temperature oscillations at the wall surface and time-shifted the temperature peak. Cabeza et al. [43] constructed outdoor cubicles in Lleida, Spain, made with plain brick, plain brick with polyurethane insulation and 1.9 mass% macro-encapsulated

PCM with a melting temperature of 27 °C, alveolar brick and alveolar brick containing 3.3 mass% macro-encapsulated PCM with a melting temperature of 25 °C. The cubicles were equipped with a heat pump to maintain an indoor temperature of 24 °C. Their electricity consumption over the course of a summer week was reduced by up to 15 and 17% by adding PCM to the plain and alveolar brick cubicles, respectively. However, as acknowledged by the authors, the performance could be further improved by optimizing the PCM melting temperature. Such optimization would be costly and time-consuming to perform experimentally. Moreover, it remains unclear whether these conclusions would be valid in other parts of the world with different climates. Rigorous numerical simulations can address these issues by assessing the effects of the climate conditions and of design parameters of PCM composite walls such as the melting temperature and the PCM volume fraction on the thermal load of buildings in a rapid, systematic and rational way.

3.5 Trombe Walls

The concept of Trombe wall was patented by E.S. Morse in the nineteenth century and developed and popularized in 1957 by Félix Trombe and Michel. In 1967, in Odeillo, France, they built the first house using a Trombe wall [44]. Figure 3.3 illustrates the general classification. Each configuration of Trombe wall is discussed in detail next.

Some examples of heating-based type of Trombe wall are (i) the photovoltaic (PV) Trombe wall, which was invented by incorporating solar cells with classic Trombe wall. The PV-Trombe wall not only provides space heating, but also generates electricity; meanwhile it brings more aesthetic value; (ii) the cooling-based type of Trombe wall, i.e. the ceramic evaporative cooing wall. The wall employs an external reflective thermal insulation blinds to avoid direct solar gain.

A Trombe wall is a primary example of an indirect-gain approach. A single or double layer of glass or plastic glazing is mounted about four inches in front of the wall's surface. Solar heat is collected in the space between the wall and the glazing. The outside surface of the wall is of black colour that absorbs heat, which is then stored in the wall's mass. Heat is distributed from the Trombe wall to the house over

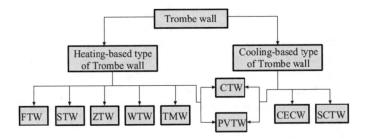

Fig. 3.3 Trombe wall classification [44]

Fig. 3.4 Schematic diagram
of PCM Trombe wall [10]

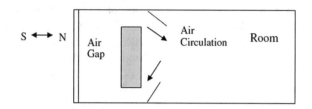

a period of several hours. When the indoor temperature falls under that of the wall's surface, heat begins to radiate into the room. Heat loss from the Trombe wall can be controlled by an insulating curtain that is closed at night in the space between the glazing and the wall. Traditionally, Trombe walls relay on sensible heat storage, but because of the potential for greater heat storage per unit mass, the PCM Trombe wall is an attractive concept still awaiting successful implementation [10]. Schematic diagram of PCM Trombe wall is shown in Fig. 3.4.

Over time, modifications have been made to Trombe walls in order to improve their efficiency. Based on the main utilizing functions of Trombe walls, they are classified into two types: a heating-based type of Trombe wall and a cooling-based type of Trombe wall. To increase the thermal resistance of the classic Trombe wall and control supplies, another heating-based type of Trombe wall, which is known as composite Trombe wall or Trombe–Michel wall, was developed.

During the last decades, several modifications have been developed from the basic design of a classical Trombe wall and composite Trombe–Michel wall [45–48]. Zalewski et al. [49] performed an experimental study of a small-scale composite solar wall where PCM was inserted into the wall in the form of brick-shaped package. The PCM used is a mixture of hydrated salts (water + $CaCl_2$ + KCl + additives) with melting point of 27 °C. They concluded that the solar gains are released with a time lag which indicates the advantage of this composite solar wall. They also pointed out that the efficiency of the solar wall could be improved by limiting losses to the outside and increasing exchanges in the cavity.

The Trombe–Michel solar wall or composite solar wall is a variant of the classic Trombe wall with the aim to contribute to the thermal comfort in buildings using solar energy. The main idea of the Trombe–Michel wall is to use the component as a part of building envelope that stores energy from solar radiation during the day and releases it as heat, indoor at night or in a cloudy period [50]. The thermal energy recovered by this type of wall passes through the storage element by conduction transfer and then by natural convection in a ventilated air gap.

In conventional non-PCM applications, the storage capacity increases weight and volume of passive solar systems, which makes difficult their merge with, common today, lightweight construction methods and limits integration into the existing buildings. To alleviate this problem, conventional heavyweight thermal mass is replaced by PCM. For a given amount of heat storage, the phase change units require less space than water walls or mass Trombe walls and are much lighter in weight. These are, therefore, much convenient to make use of in retrofit applications of buildings. Salt hydrates and hydrocarbons were used as PCMs in the Trombe wall [10].

A large number of experimental and theoretical assessments have been conducted to investigate the energy performance and long-term reliability of the PCM-based Trombe wall heat storage components. It was found that the thermal performances of the Trombe wall depend on various parameters such as the size of air gap and vents, wall area and orientation, wall thickness, glazing, insulation and operation strategy [44]. In terms of the selection of Trombe wall materials, a study was carried out by Stazi et al. [51] during both the pre-use phase and use phase of Trombe wall on three wall materials: concrete, brick and aerated concrete. Considering both pre-use and use phases, the best overall performance was obtained using the wall with aerated concrete blocks that combines a production cycle with low environmental impacts and high energy performances in the use phase.

Onishi et al. [52] numerically investigated the effects of PCM as a heat storing material on the performance of a hybrid heating system with a computational fluid dynamics (CFD) code. Simulated results indicated the effectiveness of PCM and suggested the possibility of developing low-energy houses with the hybrid system introduced in this study. The transparent insulation materials (TIM)–phase change material (PCM) wall system also showed the higher efficiency of solar radiation utilization and decreased heat losses by using corresponding PCMs [53, 54].

So, initially, hydrated slats have been sampled for this purpose. Telkes [55–57] proposed the inclusion of PCMs in walls, partitions, ceilings and floors to serve as temperature regulators. The PCMs have been used to replace masonry in a Trombe wall and worked on a construction similar to the Trombe wall, using Glauber's salt behind a polyhedral glazing. Her work was only a first-order theoretical analysis demonstrating the potential for energy and space savings.

Askew [58] used a collector panel made of a thin slab of paraffin wax and mounted behind the double glazing of the building and found that the thermal efficiencies are comparable with the conventional flat plate collectors. Farouk and Guceri [11] studied the usefulness of the PCM wall installed in a building for night-time home heating using Glauber's salt mixture and SUNOCO P-116 wax. It was observed that if the PCM wall is designed properly, it eliminates some of the undesirable features of the masonry walls with comparable results.

In experiments performed by Swet [59], Ghoneim et al. [60] and Chandra et al. [61], a Glauber's salt was utilized as well (sodium sulphate decahydrate with melting point 32.1 °C) as a phase change material in a south-facing Trombe wall. Experimental and theoretical tests were conducted to investigate the reliability of PCMs as a Trombe wall. They reported that Trombe wall with PCM of smaller thickness was more desirable in comparison to an ordinary masonry wall for providing efficient thermal energy storage (TES).

Knowles [62] presented numerical results as well as approximate simple stationary state formula with the purpose of establishing guidelines for the design of low-mass, high-efficiency walls. One conclusion was that thermal resistance of the wall should be as low as possible. Exploration of binary and ternary composite of metals, masonry and phase change materials was studied. Compared with concrete, paraffin–metal mixtures were found to offer a 90% reduction in storage mass and a 20% increase in efficiency.

Bourdeau and Jaffrin [63] and Bourdeau et al. [64] simulated and tested a Trombe wall using chliarolithe ($CaCl_2 \cdot 6H_2O$) as a PCM heat storage. A numerical model demonstrated that a 3.5-cm wall using PCM could replace a 15-cm-thick conventional wall made of concrete. In a following project, Bourdeau [65] studied the behaviour of Trombe wall made of polyethylene containers placed on a wood shelf behind a double glazing. These experimental results were used to validate the numerical model, which demonstrated that a Trombe wall with latent heat storage was more efficient than conventional concrete walls. This research indicated that the optimum thickness of a PCM wall was of a factor 4 thinner than an equivalent concrete wall.

In addition, Benson et al. [66] carried an analysis on polyalcohols used as PCM. They also performed numerical analysis on PCM-enhanced Trombe walls compared to conventional concrete structures. They found an optimum melting temperature for PCM which was close to 27 °C. Numerical analysis demonstrated that an increase in thermal diffusivity can be beneficial to the thermal performances of PCM solar walls. Accordingly, laboratory tests demonstrated that diffusivity can be increased by a factor of five through the addition of 2% of graphite, which should lead to about 30% improvement in performance. They concluded that a Trombe wall containing PCM could be four times thinner and a factor nine lighter than its equivalence made of concrete.

Buddhi and Sharma [67] measured the transmittance of solar radiation through phase change material at different temperatures and thickness. Stearic acid was used as a phase change material. They found that transmittance of the phase change material was more than the glass for the same thickness and suggested a new application of phase change material in windows/walls as a transparent insulating material.

Stritih and Novak [68] presented a solar wall for building ventilation, which absorb solar energy into black paraffin wax (melting point, 25–30 °C). The stored heat was used for heating the air for the ventilation of the house. The efficiency of the absorption was found to be 79%. The result of the simulation showed that the panel dictates the amount of stored heat as sensible or latent and that the melting point of the PCM has an influence on the output air temperature. The analysis for the heating season gave the optimum thickness of 50 mm and the melting point a few degrees above the room temperature.

Sun and Wang [69] studied the energy-saving characteristics in winter, using an experimental room. They explored a new system: passive solar collector–storage wall contained PCMs on both sides surface. The heat transfer performance and energy-saving characteristics were investigated theoretically and experimentally. Paraffin/expanded perlite/graphite PCMs was added into collector mortar layer and interior mortar layer to storage energy. Phase change temperature and latent heat are 19.45 °C and 128.46 J/g, respectively. A part of solar energy transmits into the room through air channel and vents to improve indoor temperature, a part of solar energy store in the collector mortar layer through PCMs and the rest solar energy is conducted slowly through massive wall to the room by radiation and convection. The advantage of passive solar collector–storage wall system with PCMs is storing more heat from sun during the day and releasing it into the building during the night. The results indicate that the new passive solar collector–storage wall system with PCMs can promote indoor air thermal circulation and decrease indoor air temperature fluctuations. Its good heat storage capacity can apparently improve indoor thermal environment.

Fiorito [70] selected five cities of different climate zones in Australia and modelled the effect of PCMs (n-paraffin and wax) integrated in collector–storage walls. The simulation results showed that PCM improved the thermal inertia of lightweight constructions and its position and melting temperature need to be optimized according to the corresponding climate conditions.

Kara and Kurnuc [71] applied novel triple glazing for PCM wall (33 wt% paraffin granules in the plasterboard) to prevent overheating in summer, and their experimental results indicated that the wall including PCM (GR35) with relatively lower melting temperature $t_m = 34$ °C presents better performance than that including PCM (GR41) with $t_m = 45$ °C while both could provide 14% of annual heat load of the test room.

Li and Liu [72] experimentally investigated the thermal performance of a PCM (paraffin, $t_m = 41$ °C)-based solar chimney under three different heat fluxes on the absorber surface and found that 700 W/m^2 of heat flux drives the highest air flow rate (0.04 kg/s) while 500 W/m^2 generates the highest average outlet temperature (20.5 °C). They also reported that phase change periods are nearly 13 h:50 min for all cases investigated.

In Zhou et al. [44] a test was carried out for a whole day with charging period of 6.5 h and discharging period of 17.5 h, respectively. Wall and air temperatures as well as air velocity in the gap were measured for analysis. The results showed that the PCM surface temperature increases first rapidly, then slowly and rapidly again during the charging process, which in turn corresponds with the three storage stages: sensible heat (solid), latent heat (melting) and sensible heat (liquid), respectively; the indoor temperature was found to be above 22 °C during the whole discharging period (17.5 h) under present conditions, which indicates that the indoor thermal comfort could be kept for a long time by using PCM in collector–storage wall system.

Hu et al. [73] achieves the main conclusion, the larger Trombe wall area means the more high efficiency. However, it is limited the total south wall area, that is, it is related to (α), the ratio of the Trombe wall's area to the total wall area. Massive wall materials and thickness contribute importantly to the efficiency of the wall's heat storage and release capacity. Any material characterized by high storage capacity can be used to construct Trombe walls. However, the use of lightweight materials with high storage capacity in a relatively small volume is more preferable, such as PCM. With regard to the wall thickness, 30–40-cm concrete Trombe walls have performed well in many geographical locations. In addition, thermal insulation on massive wall is considered a remedy for the deficiency of a classic Trombe wall. Glazing properties, such as the materials and the number of glazing layers significantly affect the performance of Trombe walls.

However, the selection of glazing depends on many variables including the longitude and latitude of the project. Normally, low double glazing is recommended.

Channel depth mainly contributes to the flow resistance. It is not only related to height of Trombe wall but also depend on the dimension of inlet and outlet. In addition, the structural safety should be considered when design the channel depth because an excessive depth of channel will results in an insufficient thickness of massive wall. Shading devices, such as overhang, roller shutter and venetian blinds, can control the performance of Trombe wall and address some of the shortcomings: overheating in hot summer and heat loss in winter night. Similar to thermal insulation on massive wall, proper insulation of building envelope has performed well.

Due to solar radiation can strike the indoor floor or its adjacent walls directly through a window, the design of windows should be considered including the size and position (relative to the Trombe wall). Solar radiation level has an important influence in generating air movement in a Trombe wall channel.

Generally, Trombe wall efficiency increases with increasing of solar radiation. Moreover, for a building with Trombe walls located in the north hemisphere, the south-facing facade (with 45° variations) seems to be the most effective orientation in capturing the solar gain. The wind speed and direction are related to the heat loss coefficient and wind pressure. The Trombe wall tends to perform better if the wind speed is small, and in this direction, further investigation should be carried out in the future.

3.6   Windows

The glazed areas and the shading devices have a significant role over the energy building consumption, and so many research studies and prototypes have been developed in the last years to increase the thermal and the energy efficiency of this boundary. The improvement of the thermal performance through the glazing area of the building can be accomplished resourcing to new materials, geometries and new techniques to produce solutions with higher energy efficiency [74]. New approaches, as the building orientation and the use of natural resources, as wind and solar radiation could decrease the energy needs and improve the energy transfer of these boundaries.

From the thermal perspective, PCM windows work like the optically transparent or translucent Trombe walls. They usually consist of a single or multilayer glazing panel made of conventional glass, integrated with a layer of a transparent or translucent PCM product. Figure 3.5 shows the different configuration options of semi-transparent PCM solar fenestration.

Nowadays, the building design frequently includes large translucent areas, mainly in offices and commercial buildings. However, the use of large facade glazing areas could lead to thermal and visual discomfort of indoor space and their occupants [75]. Windows represent a part of the building that is considered to lead to higher energy consumption. In warm climates dominated by cooling loads, excessive solar heat gain lead to an increased need for mechanical cooling. In cold climates, large parts of the energy escape through glazed facades, leading to a need for mechanical heating [76].

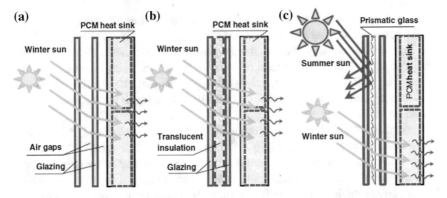

Fig. 3.5 Different configuration options of semi-transparent PCM solar fenestration: **a** semi-transparent Trombe wall containing PCM heat sink; **b** translucent Trombe wall with PCM heat sink and translucent insulation; **c** solar fenestration system using PCM heat sink and selective prismatic glass

However, glazed facades still suffer from low thermal inertia and have no way of storing excess heat. Transparent PCMs for use in windows represent an opportunity that has been explored for this purpose. They provide dynamic thermal characteristics and a source of natural lighting to the building. The energy state of these assemblies is visualized as transparent or translucent when PCM is melted and milky when PCM is frozen [4].

Manz et al. [54] studied a solar facade composed of transparent insulation material and translucent PCM used both for solar heat storage and daylighting. The PCM was hexahydrated calcium chloride ($CaCl_2 \cdot 6H_2O$) with 5% of additives. The numerical model was developed for analysis of the radiative heat transfer inside the PCM-enhanced solar window. The authors concluded that overall system performance could be improved by changing of the PCM melting temperature from 26.5 to about 21 °C.

Another semi-transparent solar window system containing PCM has been introduced by the INGLAS company form Friedrichshafen, Germany [77]. This technology combines design principles of passive solar walls with fenestration function and a semi-transparent heat reservoir. As a result, this solar window efficiently transfers solar light and absorbs the heat developed in the process. The absorbed heat is stored by the heat sink utilizing organic PCM. According to manufacture, large amounts of solar energy can be stored during daytime and released into the building at night, when PCM cools down and solidifies.

In similar research performed in Germany, an application of semi-transparent PCM components from Dorken has been jointly investigated by the glass company Glaswerke Arnold and research institute ZAE Bayern. A complete system is made of two glass sheets on the outside and a macro-encapsulated PCM on the inside [4]. As of today, only translucent PCMs have been used for PCM windows, though they enable relatively high amounts of visible light to pass through, they do not offer the same visibility as regular windows. PCM optical properties are changing to some degree between the solid and liquid states.

The solar transmittance of a commercial grade PCM was tested by Jain and Sharma [78]. For a pure PCM with a thickness of 4–30 mm, the solar transmittance was found to be 90.7–80.3%, respectively. Due to the fact that PCMs have low thermal conductivity, they concluded that PCMs may be interesting as a transparent thermal insulating medium.

Goia et al. [79] compared a prototype PCM glazing system (DGU_PCM) with a conventional double glazed (DGU_CG), insulating unit with regard to the effect on thermal comfort in the winter, summer and mid-seasons. The two glazing systems were measured over a six-month experimental campaign, and the data was used to numerically evaluate an indoor office room. The thermal conditions were considerably improved for most parts of the year with the PCM prototype compared to the regular double glazing. However, on cloudy days, the two systems showed similar properties on thermal comfort. The study also highlighted the importance of selecting the correct melting temperature for the PCM as this could affect the system negatively if not optimized. Two layers (8 mm and 6 mm) of glass compose the double glazed unit with PCM (DGU_PCM) and the air cavity (15 mm) between the two layers was filled with commercial paraffin (RT35). The DGU_PCM element area is about 1 m^2 (1.4 × 0.72 m^2), inserted in an experimental test cell with 3.6 m wide, 5.4 long and 3.0 high. The PCM was introduced into the air cavity and the volume of the PCM was 13 dm3 (E13 kg of RT35). They performed a differential scanning calorimetry (DSC) analysis to characterize the thermal properties of the PCM. According the DSC, analysing the latent heat of fusion (Δh) is about 145 kJ/kg and the melting temperature about 34 °C with a temperature range of 10 °C. They concluded that the indoor conditions reached by the application of the DGU_PCM solution increased considerably the thermal comfort comparatively to the conventional solution (DGU_CG) for the most time of the different seasons. They suggest (i) the application of PCM with lower melting temperature for cold climates and winter season and (ii) the application of PCM with high melting temperature for hot climates and summer [79].

A similar study was performed by Weinläder et al. [80], comparing a double glazing unit combined with PCM compared to a regular double glazing unit. The test specimens were studied experimentally in an outdoor test facility, and the data gained from the experiment was used for further simulations. The light transmittance from this solution was found to be 0.4, giving them the ability to be used as day-lighting elements. Compared to regular double glazing, they gave a more equalized energy balance, achieving moderate heat gains with very low heat losses. During the winter season, PCM windows improved the thermal comfort considerably and in the summer, they shifted heat gains to later in the evening. However, the PCM windows cannot be used where visual contact to the environment is desired.

Goia et al. [81] performed a full-scale test on a PCM glazing prototype. The test was performed on a south-facing wall during summer, mid-season and winter days in a sub-continental climate and compared to a conventional double glazing for reference. The experiment showed that PCM glazing can reduce the energy gain in the summer by more than 50%. In the winter, heat loss reduction during the day was observed; however, this technology might not be effective if the purpose is to achieve solar heat gains. The experiment also pointed out the importance of removing the stored heat during the night via, e.g. night cooling, and selecting the correct phase change temperature, if the system is to function optimally.

Grynning et al. [82] performed measurements on a state-of-the art commercial glazing including a PCM in a large-scale climate simulator. The tested glazing was a 4-layer glazing with a prismatic glass in the outer pane, and a PCM fill in the cavity between the inner panes. The study showed that characterization of static components (e.g. U-value, solar heat-gain coefficient, thermal transmittance) is insufficient for describing the performance of PCMs due to their dynamic nature.

A zero-energy office building using translucent PCMs in the window construction has been built in Kempen, Switzerland. In this project, each second window panel has been equipped with PCM windows. The aim for the windows is to effectively store solar energy during the warmer parts of the day, and release thermal energy throughout the colder periods of the day to reduce the total energy required for space heating [83].

Alawadhi [84] investigated the possibility to implement PCMs in window shutters, where the goal was to see if the solar heat could be absorbed before it reached the indoor space. The results indicated that the magnitude of PCM melting temperature and its quantity in the shutter have a significant effect on the thermal performance of the PCM shutter. The melting point of the PCM in the shutter should be close to the upper temperature limit of the windows during the daytime. The PCM should be prevented from completely melting during the working hours, and its amount should be sufficient to absorb large amount of heat during the daytime. When compared to foam shutters, the shutters containing a PCM could lead to a reduction in heat gain through the windows by as much as 23.29%.

In this concept, shutter-containing PCM is placed outside of window areas. During daytime, they are opened to the outside the exterior side is exposed to solar radiation, heat is absorbed and PCM melts. At night we close the shutter, slide the windows and heat from the PCM radiates into the rooms. Buddhi et al. [85] studied the thermal performance of a test cell ($1 \times 1 \times 1$ m^3) with and without phase change material. CG lauric acid (melting point, 49 °C) was used as a latent heat storage material. He found that the heat storing capacity of the cell due to the presence of PCM increases up to 4 °C for 4–5 h, which was used during night-time [10].

Mehling [86] recommended that the maximum shading temperature be delayed by 3 h and room temperature be reduced by 2 °C with the application of the PCM shutter.

Soares et al. [87] evaluated the potential of a PCM shutter-containing phase change materials. In their study, numerical studies were carried out to analyse the influence of the imposed external and internal conditions and the heat transfer coefficients of the system performance.

Li et al. [88] conducted an experiment to investigate the dynamic thermal and energy-saving performance of the experimental systems of two reference windows [DW + PCM and triple-pane window (TW + PCM)]. The experiment is conducted in summer sunny days and summer rainy days in hot summer and cold winter area of China. The result of the dynamic thermal performance of the TW + PCM (triple-pane window) in the sunny and rainy summer days shows that the TW + PCM have a good performance on reducing the temperature fluctuation indoors and saving the energy consumption.

In the sunny summer day with high outdoor air temperature and strong solar radiation, the temperature on the interior surface of the TW + PCM is 2.7 and 5.5 °C lower than that of the DW + PCM and TW, respectively, which means the overheating risk is avoided effectively, and the heat transferred into room through the TW + PCM is reduced by 16.6 and 28% compared with the DW + PCM and TW, respectively. During the discharge period of PCM in the night, the heat released from the PCM into the testing chamber reduces. It shows the same function with the DW + PCM on the delay of peak temperature time. It can adjust the peak load of the air-conditioning system and save the air-conditioning energy consumption.

In the rainy summer day with low outdoor air temperature and weak solar radiation, the temperature on the interior surface of the TW + PCM is 0.74 and 1.65°C lower than that of the DW + PCM and TW, respectively, and the heat transferred into room through the TW + PCM is reduced by 14.7% and increased by 4.5% comparing with the DW + PCM and TW, respectively. The TW + PCM play a good performance on reducing the temperature fluctuation of the interior surface and the heat entered the room, but it is unsatisfactory in reducing the peak heat flux of the interior surface and delaying the peak temperature.

Changyu et al. [89] developed a mathematical model for the heat transfer that evaluated with good agreement the effect of (i) the glass properties; (ii) the convection heat transfer coefficient; (iii) the surfaces temperatures; and (iv) the air temperature. The main results show that the influence of the solar radiation through the glass is significant for the heat flux increment. If the difference of temperatures between the surface and the air (from the inside to the outside of the room) do not exceed 2 °C, then the total heat flux effect can be neglected. The convection heat transfer coefficient does not affect significantly the overall heat flux. Heat transfer is improved by the solar radiation in winter but has an adverse effect in summer.

Gomes et al. [90] determined both solar and visible properties of a glazing system with venetian blinds using a net radiation method. They compared the numerical results with the experimental data measured from an outdoor test cell. They concluded that the developed model can be used to simulate different system configurations, such as glazing and/or shading devices (including venetian blinds) and the model can be introduced into building energy simulations and building design tools. The numerical results were compared with the experimental data and for overcast sky conditions. For these conditions, they got excellent fitting results, but for the clear sky conditions the comparison of the results presented some discrepancies. A venetian blind control strategy was used, and they conclude that is more important for southern European regions. To help designers and users to improve the thermal and daylighting indoor conditions they presented some design plots with information about how to adjust the slat orientation of the venetian blinds.

Silva et al. [74, 91–94] developed an experimental procedure that follows a test on a cell that was located in Aveiro, Portugal, and this region has a Mediterranean climate. Considering the weather database available (from the weather station at the University of Aveiro) the adequate melting temperature of the PCM was chose. According to the selected PCM the shape of the window blade (used as a macro-capsule) was defined and also the material. With the components of the window shutter prepared the system was assembled and applied in the outdoor test cell. Two compartments compose the test cell, and each has two window shutters installed (two filled with PCM for one compartment and the others two without PCM). The performance of a window shutter with phase change materials was experimentally tested at fully scale. The PCM recorded to provide an extra thermal inertia for this compartment that is recorded and pursed by the indoor air temperatures and heat flux results attained.

The maximum indoor temperature of the compartment with the PCM shutter is 37.2°C which is 16.6°C lower than the indoor air temperatures of the reference compartment. Comparing the indoor temperatures over time, the reduction of temperature can reach 90% (when the indoor air temperatures increases) and up to 35% (when the indoor air temperatures drops). For the maximum indoor air temperature peak, the difference between both compartments is 30–40%. However, the minimum indoor air temperature peaks are similar, so for this situation the potential improvement of the PCM incorporation is null, proving that other features come into play such as thermal bridges losses, large glazing areas. Therefore, an exercise of the compatibility and the optimization of these different features are necessary.

The maximum and minimum heat flux measured on the internal partition wall of the reference compartment was 16 and -8 W/m^2, respectively. The compartment with the PCM shutter presents more steady results of the heat flux and the maximum heat flux recorded was 6.5 W/m^2 and the minimum -3 W/m^2.

3.7 Concrete or Brick

Adding PCMs directly into concrete has shown some promising results through lower thermal conductivity and an increase in thermal mass at specific temperatures. However, PCM concrete has shown some undesirable properties such as lower strength, uncertain long-term stability and lower fire resistance [43].

Several studies have been conducted on PCM concrete and have shown positive effect through reduced indoor temperatures in warm climates [83]. Then, two typical construction materials, conventional and alveolar brick, were experimentally tested by Castell et al. [95].

The free-floating experiments showed that the PCM was reduced the peak temperatures up to 1 °C and smoothed out the daily fluctuations. Moreover, the electrical energy consumption was reduced about 15% in the PCM cubicles in summer. These energy savings resulted in a reduction of the CO_2 emissions about 1–1.5 kg/year/m^2. Later, to control and shape the PCMs, a new composite construction material that embedded micro-encapsulated PCM in plastering mortar was developed and tested by Sá et al. [96], indicating the peak temperature of the indoor air was reduced by 2.6 °C after the PCM mortar used. That means, when the PCMs are used in plastering mortar, concrete, brick and cubicle, the thermal environment control performance could be improved, as well as the structure stability of buildings unaffected.

Combining concrete structures with PCMs have been tried in several different ways. One studied solution is to drill holes in the concrete which may then be filled with a PCM [97].

Royon et al. [98] tested the possibility of filling the already hollow areas in a hollow concrete floor with PCMs. The concrete was filled with paraffin PCM with a melting temperature of 27.5 °C. This test showed that the temperature on the other side of the hollow concrete was lower during summer conditions. Hence, such floors can be used as a passive thermal conditioner during the summer. However, more tests are needed with real-life climate conditions to validate the effects.

A most surprising research was aimed at developing and testing the multifunctional performance of new PCM-filled structural concretes for building applications by Cabeza et al. [43]. Such concretes were prepared by adding 5% in weight of encapsulated phase change material for thermal energy storage applications. Moreover, the included PCMs were selected in order to identify the best capsule geometry for structural and thermal purpose. In particular, the same paraffin PCM with melting temperature at 18 °C was selected and industrially capsulated in two ways. The micro-capsules included about 85–90% of PCM in small capsules with diameter of about 17–20 μm. The second configuration concerned the macro-capsuled PCM, consisting in a sort of matrix with PCM micro-capsules inside, having a whole diameter of 3–5 mm and a whole PCM concentration of 80%.

3.8 Underfloor Heating

Areas which are in direct contact with solar radiation hold large potential for storage of thermal heat energy. Floor solutions incorporating PCMs in areas of a building where the sun shines for large parts of the day may benefit from incorporating PCMs. Figure 3.6 shows flat profiles filled with PCMs that were used in the floor in North House, a competitor in the US Department of Energy's solar decathlon.

Xu et al. [99] performed a simulation of the thermal performance of PCMs used in a passive floor system during the winter season. The performance on the systems was influenced by the choice of covering material, the air gap between the PCM and covering material and the thickness of the PCM. For the simulations performed, the thickness should not exceed 20 mm as this would not increase the influence of the thermal storage significantly. As reported, the application of PCMs in concrete floors resulted in a reduction of maximum floor temperatures up to $16 \pm 2\%$ and an increase of minimum temperatures up to $7 \pm 3\%$ [100].

A hollow concrete floor panel was incorporated with PCMs by Royon et al. [101, 102]. Thermal response to a temperature variation was also investigated, showing a

Fig. 3.6 Flat profiles which can be installed underfloor to store and release latent thermal heat energy

decrease of the surface wall temperature amplitude and an increase of thermal energy stored for this novel floor. Clearly, it is very convenient for this method to storage energy in the PCMs.

Ansuini et al. [103] using a lightweight piped radiant floor prototype with an integrated PCM layer, with their experimental floor specimen. A new PCM floor was also investigated by Huang et al. [104], revealing the new PCM floor was able to release 37,677.6 kJ heat for 16 h in the pump-off period in a room of 11.02 m^2 and that accounted for 47.7% of energy supplied by solar water.

Later, the performance of a PCM floor radiation heating system was experimentally investigated by Zhou et al. [105], indicating the advantages of using PCM-capillary mat combination for low-temperature floor panel typical of solar-hot-water heating system. A two-dimensional coupled heat transfer model based on variable thermophysical parameters of PCM was established by Zhao et al. [106], concluding that the air temperature fluctuation in the cavity with PCM structure was in a smaller magnitude.

A new double-layer radiant floor system with organic PCMs was proposed and tested by Xia et al. [107], showing that the double-layer radiant floor system with PCM could meet the thermal need of users under heating mode. The above research studies showed that the designed PCM floor was capable of achieving large-span intermittent heating and lower thermal conductivity for the decoration material, and helpful for adjusting the floor surface temperature in the present design.

PCM-embedded floor and a chilled ceiling also attracted researchers' attention. Belmonte et al. [108] reported a numerical study on PCM incorporated into the floor, and a hydronic radiant ceiling system was used as the energy discharge system. The simulation results revealed that when accompanied by an air-to-air heat recovery system, this configuration could reduce the cooling energy demand and the energy consumption more than 50%. However, the degrees of occupant comfort will inevitably vary, for example the predicted percentage dissatisfied (PPD) increases by 2–5%. In the view of thermal comfort study, this system is valuable to be further investigated.

In summary, studies on PCM floor are concentrated on two aspects, (1) integrating PCMs into floor materials or PCM slabs as a layer in the floor construction, and (2) coupling a PCM-TES with floor heating system. All researchers payed their attention on energy saving for building, while few of them take the indoor thermal comfort into consideration in their experimental and simulation studies. Although most of experimental cases on PCMs were carried out, most of them were tested in the laboratory. To exactly calculate the indoor thermal comfort, a true building environment should be prepared, and thus field test in real building should be conducted.

3.9 Ceilings

Implementing PCMs into roof systems does not seem to have received much attention, a few studies on the possible effects of PCMs in passive roof systems have been found. The thought is that PCMs placed on the roof will be able to absorb the incoming solar energy and the thermal energy from the surroundings to reduce temperature fluctuations on the inside [83].

Pasupathy and Velraj [109] studied the effects of a double layer of PCM for year round thermal management in Chennai, India. An experiment was performed with a PCM roof panel compared to a reference room without the PCM panel. The PCM used was an inorganic eutectic of hydrated salts. The experiment showed that the PCM panel on the roof narrowed the indoor air temperature swings, and that such a system could perform during all seasons when the top panel had a melting temperature 6–7 °C higher than the ambient temperature in the early morning during the peak summer month, and the bottom panel had a melting temperature near the suggested indoor temperature.

Kosny et al. [110] set up a naturally ventilated roof (see Fig. 3.7) with a photovoltaic (PV) module and PCMs to work as a heat sink. The goal was that the PCM would absorb heat during the day in winter and release it in the night to reduce heating loads. In the summer, the PCM would absorb heat to reduce the cooling loads in the attic beneath. A full-scale experiment was performed over a whole year from November 2009 until October 2010 in Oak Ridge, Tennessee.

The data from the tests were compared with a conventional asphalt shingle roof. The PV-PCM attic showed a 30% reduction in heating loads during the winter and a 55% reduction in cooling loads. Furthermore, a 90% reduction in peak daytime roof heat fluxes was observed.

According to Kalnæs and Jelle [83], the thought is that PCMs placed on the roof will be able to absorb the incoming solar energy and the thermal energy from the surroundings to reduce temperature fluctuations on the inside.

Koschenz [111] developed a study that describes the development of a thermally activated ceiling panel for incorporation in lightweight and retrofitted buildings. As alteration and refurbishment schemes look set to account for an increasing propor-

Fig. 3.7 PV-PCM roof [110]

tion of construction work, the focus was on minimizing overall panel thickness while providing ample storage capacity. It was demonstrated, by means of simulation calculations and laboratory tests that a 5-cm layer of micro-encapsulated PCM (25% by weight) and gypsum suffice to maintain a comfortable room temperature in standard office buildings. The system's features also make it ideal for use in lightweight structures, the incorporation of additional thermal mass offering an efficient means of moderating temperature amplitudes in this type of building.

Also in this, we can found that there are some simulation calculations for specification of required ceiling panel properties. Simulation calculations were performed to determine the required ceiling panel characteristics, based on the properties of the basic materials. Key parameters included the thickness of the PCM/gypsum composite layer, the proportion of paraffin and the minimum requirements placed on the PCM in terms of melting range and latent heat of fusion. An overall panel thickness of 5 cm is required to store the total heat gain of 320 Wh/m^2 day. The quantity of PCM in the gypsum must be at least 25% by weight. In order to meet the required temperature boundary conditions, it is important for the melting range of the paraffin to be carefully adjusted to the specific situation. The region with maximum values of specific heat must therefore correspond to 21–22 °C.

To avoid large temperature gradients inside the material, the ceiling panels must exhibit good thermal conductivity over the entire cross section. Here, the simulation calculations showed a mean target value of $\lambda = 1.2$ W/mK to be practicable.

Weinläder et al. [112] carried out a research that was part of the project "Development and Practical Performance Testing of Building Components with PCM in Demonstration Buildings". In this paper, a ventilated cooling ceiling with PCM was evaluated in a long-term monitoring programme in two offices and a conference room. The system showed a significant cooling potential in summer and some considerable synergetic effects if combined with a sun protection system with PCM. While the ventilated cooling ceiling with PCM and the sun protection system with PCM alone reduced the maximum operative room temperature by 2 K, the temperature reduction with both systems together was up to 4 K. The ventilation system of the cooling ceiling would also be advantageous for the regeneration of the PCM blinds.

3.10   Thermal Insulation Materials

Recently, incorporation of PCMs into fibrous thermal insulation materials has received considerable attention. Kosny et al. [113] performed an experimental and numerical analysis of a wood-frame wall containing PCM-enhanced fibre insulation. The wall assembly had an R-value of 4.14 m^2K/W (a U-value of 0.241 W/m^2K). For fibre insulation filled with 30 wt% PCM in summer conditions, results showed a reduced peak hour heat gain of 23–37% in Marseille and 21–25% in Cairo and Warsaw.

Lei et al. [114] developed a numerical simulation in order to calculate building energy performance by integrating a PCM for cooling load reduction in a tropical climate using Energy Plus. The results demonstrated that a small amount of heat-gain reduction of 4.5% was achieved with a 10 mm "solid PCM" layer. In addition, a meaningful heat-gain reduction of around 40.7% was obtained by adding another PCM layer.

Jin et al. [115] experimentally studied the influence of PCM position on the thermal performance of building walls. The effects of the state of a variety of salt-hydrate PCM on its phase change performance were investigated with differential scanning calorimeter (DSC) examinations and cooling experiments. The results demonstrated that the PCM was in the fully melted state before cooling when the PCM locations were 2/5 L (near internal layer), 3/5 L, 4/5 L and L, while in the PCM location of 1/5 L, the PCM was in the partially melted state prior to cooling.

Izquierdo-Barrientos et al. [116] developed a one-dimensional transient heat transfer model using a finite-difference method to investigate the impact of PCMs in external building walls. Furthermore, several external building wall arrangements were analysed for a typical building wall by varying the position of the PCM layer, the orientation of the wall, the ambient conditions and the phase transition temperature of the PCM. They found that there was no important reduction in the total heat lost during winter regardless of the wall orientation or PCM transition temperature.

Kuznik et al. [117] investigated an experimental method for the thermal performance of a PCM-copolymer-composite wallboard in three different climates. The results showed that for all the cases tested, the ratio between the amplitude of the indoor air temperature in the cell with PCM and the amplitude of the reference test cell was between 0.73 and 0.78. In addition, the PCM tested kept the room air temperature within the comfort zone while the maximum air temperature of the room could be decreased by as much as 4.2 °C.

Silva and Vicente [93] investigated an experimental testing of wall elements integrated by PCM macro-encapsulation. The results showed that the maximum amplitude reductions were about 50 and 80% for two different specimens. In addition, the delay time within the source specimen and the imposed temperature was nearly 1 h, but this value extended to 3 h for other specimens with the incorporation of PCM.

Evola et al. [118] reported a complete evaluation methodology for optimizing PCMs to enhance summer thermal comfort in lightweight buildings. In addition, they investigated the intensity and duration of the thermal comfort for the occupants.

Finally, they analysed a new case study, including organic PCMs, on a lightweight office building. This study was useful to assist with detection of the most appropriate PCM and its installation pattern as a function of the climatic operating conditions and of the comfort requirements.

Jin et al. [119] investigated the dependence of wall thermal performance on PCM location leading to the most appropriate PCM locations. They evaluated a dynamic model of walls with and without a PCM thermal shield. The experimental results showed that the optimum location of the PCM thermal shield was 1/5 L from the interior. In this location, the peak reduction of heat flux was approximately 41% and time lag was around 2 h. They also evaluated the optimal location of a "thin" PCM layer while varying PCM thickness, melting temperature, heat of fusion and interior surface temperatures, to increase thermal capacity and shave peaks of heat flux within the wall. Based on a mathematical model, the optimal locations of the PCM layer were determined to be closer to the exterior surface of the wall. While the thickness, heat of fusion, and the melting temperature of PCM were increased, the optimal position of PCM was closer to the interior surface.

3.11   Furniture and Indoor Appliances

A point that has not been investigated with the same affluence, but should be mentioned, is the possibility of using PCMs in furniture and other indoor appliances. The benefit of PCMs is as mentioned their ability to store heat in periods where there is a surplus and release the heat when there is a deficit. It would be interesting to study how incorporation of PCMs into other components in a building besides the structural components could benefit energy savings and thermal comfort [83].

PCMs have already been widely studied for textile applications [120], showing that there is a possibility of adding PCMs to various forms of materials. The large surface area of the furniture exposed to the indoor environment can be ingeniously used for latent heat thermal energy storage (LHTES) with the integration of phase change materials (PCMs). Their appreciable energy storage density is an interesting asset for increasing the thermal inertia of light structure buildings and for extending the applicability of the TES strategy. PCM furniture could allow integration of LHTES in low thermal inertia dwellings without the need for building renovation. Horikiri et al. [121] used computational fluid dynamics (CFD) to assess the effect of room occupancy and furniture arrangement with and without heat generation in terms of local thermal comfort. Three different configurations of furniture and occupants were compared with the empty room case. The study pointed out that addition of non-heat generating furnishing in the ventilated room can induce complicated flow re-circulations and high local air velocities around edges of the furniture. However, it has little influence on room temperature and airflow buoyancy strength, compared with that of unfurnished room case. Finally, the heat generation from the TV did not have important impact on the thermal comfort and heat transfer. Analysis of the impact of occupied room on indoor thermal comfort is carried out by three different layouts/scenarios with furniture and/or occupants (S1–S3), compared with the original empty model room layout/scenario S0. The furniture considers a cabinet (or a TV stand) with a TV at a fixed position, located at the middle of one side-wall opposite to the sofa, and two different types of sofa. A small sofa that has no armrest is located at the back wall, facing to the window wall (denoted as the layout S1) while a large sofa with armrest is located at the middle of one side-wall (denoted as the layout S2). In the layout S3, two sofas are both included. All sofas and cabinet/TV stand are attached to the walls, assuming that the gap between the walls and the non-heat generating furniture is so small that the local heat transfer and fluid pattern inside the gap space do not have significant influences on the domain of interest, i.e. the central space of the model room [121].

The humidity buffering effect of materials located in the thermal zone can reduce humidity variation. It improves thermal comfort and decreases energy consumption of the mechanical systems for humidification or dehumidification.

Yang et al. [122] conducted full-scale experiments on moisture buffering capacity of interior surface materials and impact of the presence of furniture in the interior space. The results showed that the indoor humidity variation decreased by up to 12% and the total moisture buffering potential of the room increased by up to 54.6% for a fully furnished case. The authors explained that furnishing elements present much more surface area for moisture exchange and buffering than envelope inner surfaces. Furniture materials can also hold more water vapour than interior surface ones. In addition, the variation of moisture contents of walls screened by furnishing is not always the same as in an empty room. The results also indicated that a bookshelf with books and a bed with mattress present higher moisture buffering capacity than tables, chairs and curtains.

Mortensen et al. [123] investigated the local micro-climate created by furnishing elements close to cold walls. A piece of furniture placed near a poorly insulated external wall can lead to condensation on the inner side of the building envelope. The authors used particle image velocimetry to perform a two-dimensional experimental analysis of the airflow pattern in a small air gap between a chilled wall and a closet placed next to it. Two air gap widths were tested: 25 and 50 mm. Length of legs of the furniture varied from 0 to 200 mm. The study indicated that vertical flow dominates with similar behaviour as in between vertical plates heated asymmetrically. The flow in the air gap was not fully developed, and maximum velocities were found near the cold wall. Finally, the flow rate increased when the gap was expanded or if the furniture was elevated from the floor.

Corcione et al. [124] published numerical studies showing a non-negligible decrease in the heat transfer from radiant surface systems to the furnished indoor space in comparison to an empty room case. The air and mean radiant temperature were also impacted. Fontana extended this work with experimental investigations using a small-scale test setup to look at the impact of furniture pieces with different surface areas, locations and distance from the floor. The author concluded that 40% of floor covering with different kinds of furniture can reduce the heat flux from the radiant floor to the room by 25–30%.

Pomianowski et al. [125] conducted a full-scale experiment concerning the influence of an internal obstacle on the overall heat transfer in a room when using displacement night-time ventilation. The presence of a table changed the average convective heat transfer coefficient in the test chamber and the mean heat flux at the ceiling by 3.96 and 9.84%, respectively, when applying an air change rate of $6.6\,h^{-1}$. The only noticeable drops in the temperature efficiency caused by the presence of the table were observed at low air change rates.

The studies presented above pointed out that the influence of furniture cannot be neglected when designing a radiant floor system. Surprisingly, it has been found that Fontana [126] was the only one to publish the results of an experiment investigating the impact of furniture on radiant systems. As mentioned by Le Dréau [127], further experimental researches are required to quantify the effect of furniture on the effectiveness of radiant systems.

Antonopoulos and Koronaki [128] characterized the thermal capacitance, time constant and thermal delay of typical Greek detached houses with a one-dimensional finite-difference model. The authors took into account the presence of furniture thermal mass and modelled it as an equivalent one-side wooden slab of $6\,m^2$ per m^2 of floor area and a thickness of 5 cm, which gives an internal mass density of about 180 kg per m^2 of floor area. No justification was given for the choice of this value. Solar load and internal heat gains were applied to the air node only. The results showed that the envelope, partition walls and furniture represented 78.1, 14.5 and 7.4%, respectively, of the total effective building thermal capacitance. The authors concluded that furniture/indoor mass can increase the building time constant and thermal delay by up to 40% (25% for interior wall partitions and 15% for the furnishings).

Yam et al. [129] developed a simplified building model with adiabatic envelope and no internal sun load to inspect the nonlinear coupling between internal thermal mass and natural ventilation. They found that a maximum indoor temperature phase shift of 6 h can be achieved if the fresh air is directly supplied from the outdoor environment, presenting periodic temperature variations. The authors suggested that an appropriate amount of thermal mass should be used in building passive design because further increase above an optimum point does not change the phase shift of the system. Zhou et al. [105] extended the aforementioned study by adding the envelope thermal mass into consideration. The results showed that increasing the internal thermal mass of a building with a large time constant to adjust the indoor air temperature is not an effective solution.

Wolisz et al. [130] carried out a numerical analysis on the impact of modelling furniture and floor covering in thermal building simulations with temperature set point modulation control. The study cases were a massive building and a light frame building, both with very good insulation levels and underfloor heating systems. The furniture element was represented by an equivalent horizontal board of wood or metal. Long-wave radiation heat exchanges were modelled by coupling inner surfaces to a fictive massless black body node in a star network scheme. One internal wall had 50% of its surface area covered by furniture. It was found that after 4 h of increased set point, an empty massive room was 1.2 °C warmer than the one with flooring and furniture. A fully equipped massive room can have a time delay of more than 7 h to raise its temperature by 5 °C, compared to an empty room. Furnishing and floor covers can change cool-down times by up to 2 h in the case of periodic set point control.

The floor covering presented more significant effect on the heating time than the furniture element because the underfloor radiant system was used as a heating source. However, the effect of furniture became more important for the lightweight room with periodic set point scenario. The authors concluded that both the furniture and the floor covering of a room have a distinct and significant impact on the indoor temperature for dynamic set point control.

Raftery et al. [131] performed a sensitivity analysis on the influence of furniture on the peak cooling load of a large open space multi-story office building located in San Francisco. The authors used the Energy Plus software and varied multiple parameters such as type of HVAC system, building orientation, window to wall ratio, envelope thermal inertia and amount and surface area of the internal mass element. Two different furniture models were tested: a simplified non-geometric furniture element, which is not taken into account for solar radiation and long-wave heat exchange and a new model with a geometric representation of an equivalent furniture slab located in the centre of the room, 0.5 m above the floor. With the latter, direct and diffuse solar radiation repartition can be executed accordingly with shading effect of the planar element on the floor. Long-wave radiation heat exchange can also be calculated with correct view factors. Results were presented using the median value following by the lower and upper quartiles in parentheses. The study found that internal mass can change peak cooling load by −2.28% (−5.45, −0.67%). The geometric modelling changed peak cooling load by −0.25% (−1.02, +0.23%) when compared to the non-geometric model. This geometric modelling had a larger effect in cases with high direct solar radiation and almost no effect for low solar loads. The impact was also found more important for HVAC radiant systems, which yield a surface temperature asymmetry. The thickness of the internal mass element had a relatively large impact on results. Very thin elements with a small time constant convert the solar load into a convective load quickly and can thus increase the peak cooling load. The authors concluded that the choice of modelling method is not significant compared to the uncertainty on the internal mass characteristics such as surface area, material properties, weight and thickness.

3.12 Safety Requirements

The safety requirements for materials used in buildings are crucial points for the PCMs to fulfil. As mentioned earlier, PCMs should not be toxic or flammable. However, for many organic PCMs, flammability and possible release of toxic fumes during combustion have been an issue. Solutions have been made to counter this issue, such as ignition-resistant micro-capsules for PCMs and the adding of fire retardants [83].

Hence, it is of significance that manufacturers of PCMs for building applications are required to give reliable information about the fire performance of their products. Nguyen et al. [132] reviewed the work that has been carried out to improve fire safety of PCMs. This work investigated the use of fire retardants to increase fire resistance of composite PCMs.

In conclusion, incorporating phase change materials (PCM) into a building enables a more dynamic use of energy. Due to the storage capabilities of PCMs, excess heat can be stored during warm periods and released during cold periods. It may also work the other way around, storing cold energy and using it for free cooling systems in warm periods. The benefits of using PCMs in buildings mainly revolve around a decrease in energy usage along with a peak load shifting of energy required for heating or cooling and an increase in thermal comfort by decreasing temperature fluctuations [83].

References

1. J. Kosny, T. Stovall, S. Shrestha, D. Yarbrough, Theoretical and experimental thermal performance analysis of complex thermal storage membrane containing bio-based phase-change material (PCM). Proc. Build. **XI**, 1–13 (2010)
2. B. Zalba, J.M. Marín, L.F. Cabeza, H. Mehling, Review on thermal energy storage with phase change materials, heat transfer analysis and applications. Appl. Therm. Eng. **23**, 251–283 (2003)
3. H. Mehling, L.F. Cabeza, *Heat and Cold Storage with PCM. An Up To Date Introduction into Basics and Application* (Springer, Berlin, 2008)
4. C.A.P. Santos, L. Matias, Coeficientes de Transmissão Térmica de Elementos da Envolvente dos Edifícios. ICT Informações Científicas e Técnicas, Edifícios - Ite 50, Edited by Laboratório Nacional de Engenharia Civil. LNEC, Lisboa (2007)
5. Y. Zhang, G. Zhou, K. Lin, Q. Zhang, H. Di, Application of latent heat thermal energy storage in buildings: State-of-the-art and outlook. Build. Environ. **42**, 2197–2209 (2007)
6. S. Monteiro da Silva, M. Almeida, *Using PCM to Improve Building's Thermal Performance.* 2nd International Conference on Sustainable Energy Storage, 19–21 June, Trinity College Dublin, Ireland
7. S. Scalat, D. Banu, D. Hawes, J. Paris, F. Haghighata, D. Feldman, Full scale thermal testing of latent heat storage in wallboard. Solar Energy Mater Solar Cells **44**, 49–61 (1996)
8. R.J. Kedl, T.K. Stovall, *Activities in Support of the Wax-Impregnated Wallboard Concept. US Department of Energy: Thermal Energy Storage Researches Activity Review*, New Orleans, LA, USA (1989)
9. D.A. Neeper, *Solar Buildings Research: What Are the Best Directions?* 213–219 (1986)
10. V. Tyagi, D. Buddhi, PCM thermal storage in buildings: a state of art. Renew. Sustain. Energy **11**, 1146–1166 (2007)
11. B. Farouk, S.I. Guceri. *Tromb–Michal Wall Using a Phase Change Material* (1979)
12. K. Peippo, P. Kauranen, P.D. Lund, Multicomponent PCM Wall Optimized for Passive Solar Heating. Energy Build. **17**, 259–270 (1991)
13. D. Feldman, D. Banu, D. Hawes, E. Ghanbari, Obtaining an energy storing building material by direct incorporation of an organic phase change material in gypsum wallboard. Solar Energy Materials **22**, 231–242 (1991)
14. D. Feldman, M.A. Khan, D. Banu, *Energy Storage Composite with an Organic Phase Change Material* (1989), pp. 333–341

15. D. Feldman, M. Shapiro, D. Banu, C.J. Fuks, *Fatty Acids and Their Mixtures as Phase Change Materials for Thermal Energy Storage* (1989), pp. 201–216
16. M.M. Shapiro, D. Feldman, D. Hawes, D. Banu, *PCM Thermal Storage in Wallboard* (1987), pp. 48–58
17. M.M. Shapiro, *Development of the Enthalpy Storage Materials, Mixture of Methyl Stearate and Methyl Palmitate* (1989)
18. D.W. Hawes, D. Feldman, D. Banu, Latent heat storage in building materials. Energy Build. **20**, 77–86 (1993)
19. D.A. Neeper, *Potential Benefits of Distributed PCM Thermal Storage*. Proceedings of the 14th National Passive Solar Conference, 19–22 June 1989, Denver, pp. 283–288
20. D.A. Neeper, Thermal dynamics of wallboard with latent heat storage. Sol. Energy **68**, 393–403 (2000)
21. D. Heim, J.A. Clarke, Numerical modelling and thermal simulation of PCM–gypsum composites with ESP-r. Energy Build. **36**(8), 795–805 (2004)
22. J. Paris, M. Falardeau, C. Villeneuve, Thermal storage by latent heat: a viable option for energy conservation in buildings. Energy Sources **15**, 85–93 (1993)
23. A.E. Rudd, Phase change material wallboard for distributed storage in buildings. Trans.-Am. Soc. Heating Refrigerating Air Conditioning Eng. 339–346 (1993)
24. M.W. Babich, R. Benrashid R, R.D. Mounts, DSC studies of new energy storage materials. Part 3. Thermal and flammability studies. Thermochimica Acta, 193–200 (1994)
25. D. Banu, D. Feldman, F. Haghighat, J. Paris, D. Hawes, Energy-storing wallboard: flammability tests. J. Mat. Civil Eng. **10**, 98–105 (1998)
26. F. Kuznik, J. Virgone, Experimental investigation of wallboard containing phase change material: data for validation of numerical modeling. Energy Build. **41**, 561–570 (2009)
27. A. Oliver, Thermal characterization of gypsum boards with PCM included: thermal energy storage in buildings through latent heat. Energy Build. **48**, 1–7 (2012)
28. H. Liu, B.A. Hanzim, Performance of phase change material boards under natural convection. Build. Environ. **44**, 1788–1793 (2009)
29. L. Shilei, F. Guohui, Z. Neng, D. Li, Experimental study and evaluation of latent heat storage in phase change materials wallboards. Energy Build. **39**, 1088–1091 (2007)
30. L. Shilei, F. Guohui, Z. Neng, Impact of phase change wall room on indoor thermal environment in winter. Energy Build. **38**, 18–24 (2006)
31. C. Voelker, O. Kornadt, M. Ostry, Temperature reduction due to the application of phase change materials. Energy Build. 937–944 (2008)
32. F. Kuznik, J. Virgone, K. Johannes, In-situ study of thermal comfort enhancement in a renovated building equipped with phase change material wallboard. Renew. Energy 1458–1462 (2011)
33. F. Kuznik, J. Virgone, Experimental assessment of phase change material for wall building use. Appl. Energy **86**, 2038–2046 (2009)
34. A. Athienitis, C. Liu, D. Hawes, D. Banu, D. Feldman, Investigation of the thermal performance of a passive solar test-room with wall latent heat storage. Build. Environ. 405–410 (1997)
35. P. Schossig, H.M. Henning, S. Gschwander, T. Haussmann, Microencapsulated phase-change materials integrated into construction materials. Sol. Energy Mater. Sol. Cells **89**(2–3), 297–306 (2005)
36. S.G. Jeong, S.J. Chang, W. Seunghwan, J. Lee, S. Kim, Energy performance evaluation of heat-storage gypsum board with hybrid SSPCM composite. J. Indus. Eng. Chem 237–243 (2017)
37. B. Chhugani, F. Klinker, H. Weinlaeder, M. Reim, Energetic performance of two different PCM wallboards and their regeneration behavior in office rooms. Energy Procedia **122**, 625–630 (2017)
38. M. Pomianowski, P. Heiselber, Y. Zhang, Review of thermal energy storage technologies based on PCM application in buildings. Energy Build. **67**, 56–69 (2013)

39. I. Cerón, J. Neila, M. Khayet, Experimental tile with phase change materials (PCM) for building use. Energy Build. **43**, 1869–1874 (2011)
40. D.C. Hittle, *Phase Change Materials in Floor Tiles for Thermal Energy Storage* (2002)
41. R. Novais, G. Ascensão, M.P. Seabra, J.A. Labrincha, Lightweight dense/porous PCM-ceramic tiles for indoor temperature control. Energy Build. **108**, 205–214 (2015)
42. T.C. Ling, C.S. Poon, Use of phase change materials for thermal energy storage in concrete: an overview. Construct. Build. Mat. 55–62 (2013)
43. L.F. Cabeza, C. Castellón, M. Nogués, M. Medrano, R. Leppers, O. Zubillaga, Use of microencapsulated PCM in concrete walls for energy savings. Energy Build. 113–119 (2007)
44. G. Zhou, M. Pang, Experimental investigations on the performance of a collector–storage wall system using phase change materials. Energy Convers. Manag. 178–188 (2015)
45. A.K. Sharma, N.K. Bansal, M.S. Sodha, V. Gupta, Vary-therm wall for cooling/heating of buildings in composite climate. Int. J. Energy Res. 733–739 (1989)
46. L. Zalewski, M. Chantant, S. Lassue, B. Duthoit, Experimental thermal study of a solar wall of composite type. Energy Build. 7–18 (1997)
47. L. Zalewski, S. Lassue, B. Duthoit, M. Butez, Study of solar walls, validating a simulation model. Build. Environ. 109–112 (2002)
48. J. Jie, Y. Hua, H. Wei, P. Gang, L. Jianping, J. Bin, Modeling of a novel Trombe wall with PV cells.2007. Build. Environ. 1544–1552 (2007)
49. L. Zalewski, A. Joulin, S. Lassue, Y. Dutil, D. Rousse, Experimental study of small-scale solar wall integrating phase change material. Solar Energy 208–219 (2012)
50. E. Leang, P. Tittelein, L. Zalewski, S. Lassue, Numerical study of a composite Trombe solar wall integrating microencapsulated PCM. Energy Procedia **122**, 1009–1014 (2017)
51. F. Stazi, C. Bonfigli, E. Tomassoni, C. Di Perna, P. Munafò, The effect of high thermal insulation on high thermal mass: is the dynamic behaviour of traditional envelopes in Mediterranean climates still possible? Energy Build. 367–383 (2015)
52. J. Onishi, H. Soeda, M. Mizuno, Numerical study on a low energy architecture based upon distributed heat storage system. Renew. Energy 61–66 (2001)
53. U. Stritih, P. Novak, Solar heat storage wall for building ventilation. Renew. Energy 268–271 (1996)
54. H. Manz, P.W. Egolf, P. Suter, A. Goetzberger, TIM-PCM external wall system for solar space heating and daylighting. Solar Energy 369–379 (1997)
55. Telkes M. Trombe wall with phase change storage material. 1978
56. M. Telkes, Thermal energy storage in salt hydrates. Solar Mat. Sci. 381–393 (1980)
57. Telkes M. Thermal storage for solar heating and cooling. 1975
58. G.L. Askew, *Solar Heating Utilization A Paraffin's Phase Change Material* (1978)
59. C.J. Swet, *Phase Change Storage in Passive Solar Architecture* (1980), pp 282–286
60. A.A. Ghoneim, S.A. Kllein, J.A. Duffie, Analysis of collector—storage building walls using phase change materials. Solar Energy 237–242 (1991)
61. S. Chandra, R. Kumar, S. Kaushik, S. Kaul, Thermal performance of a non-A/C building with PCM thermal storage wall. Energy Convers. Manage. 15–20 (1985)
62. T. Knowles, Proportioning composites for efficient thermal storage walls. Solar Energy 319–326 (1983)
63. L. Bourdeau, A. Jaffrin, *Actual Performance of a Latent Heat Diode Wall* (1979)
64. L. Bourdeau, A. Jaffrin, A. Moisan, Captage et Stockage d'ànergie Solaire dans l'Habitat par le Moyen de Mur Diode à Chaleur Latente 559–568 (1980)
65. L. Bourdeau, Utilisation d'un Materiau à Changement de Phase Dans un Mur Trombe sans Thermocirculation (1982), pp 633–642
66. D.K. Benson, J.D. Webb, R.W. Burrows, J.D.O. McFadden, C. Christensen (1985) *Materials Research for Passive Solar Systems: Solid State Phase-Change Materials* (1985)
67. D. Buddhi, S.D. Sharma, Measurements of transmittance of solar radiation through stearic acid: latent heat storage material. Energy Convers. Manag. 1979–1984 (1999)
68. U. Stritih, P. Novak, Solar heat storage wall for building ventilation, In: World renewable energy congress (WREC). Renew. Energy. 268–271 (1996)

69. D. Sun, L. Wang, Research on heat transfer performance of passive solar collector-storage wall system with phase change materials. Energy Build. **199**, 183–188 (2016)
70. F. Fiorito, Trombe walls for lightweight buildings in temperate and hot climates: exploring the use of phase-change materials for performances improvement. Energy Procedia 1110–1119 (2012)
71. Y.A. Kara, A. Kurnuc, Performance of coupled novel triple glass and phase change material wall in the heating season: an experimental study. Solar Energy 2432–2442 (2012)
72. Y.C. Li, S.L. Liu, Experimental study on thermal performance of a solar chimney combined with PCM. Appl. Energy **114**, 172–178 (2014)
73. Hu Z, He W, Ji J, Zhang S, Hu Z, He W, A review on the application of Trombe wall system in buildings. Renew. Sustain. Energy Rev. 976–987 (2017)
74. Silva Tiago, Vicente Romeu, Rodrigues Fernanda, Literature review on the use of phase change materials in glazing and shading solutions. Renew. Sustain. Energy Rev. **53**, 515–535 (2016)
75. F. Cappelletti, A. Prada, P. Romagnoni, A. Gasparella, Passive performance of glazed components in heating and cooling of an open-space office under controlled indoor thermal comfort. Build. Environ. 131–144 (2014)
76. K.A.R. Ismail, C.T. Salinas, J.R. Henriquez, Comparison between PCM filled glass windows and absorbing gas filled windows. Energy Build. 710–719 (2008)
77. http://www.inglas.eu/glass/company.html. Accessed 17.11.2017
78. F. Goia, M. Perino, V. Serra, Improving thermal comfort conditions by means of PCM glazing systems. Energy Build. 442–452 (2013)
79. L. Jain, S.D. Sharma, Phase change materials for day lighting and glazed insulation in buildings. J. Eng. Sci. Technol. 322–327 (2009)
80. H. Weinläder, A. Beck, J. Fricke, PCM-facade-panel for daylighting and room heating. Solar Energy 177–186 (2005)
81. F. Goia, M. Perino, V. Serra, Experimental analysis of the energy performance of a full-scale PCM glazing prototype. Solar Energy 217–233 (2014)
82. S. Grynning, F. Goia, E. Rognvik, B. Time, Possibilities for characterization of a PCM window system using large scale measurements. Int. J. Sustain. Built. Environ. 56–64 (2013)
83. S.E. Kalnæs, B.P. Jelle, Phase change materials and products for building applications: a state-of- the-art review and future research opportunities. J. Sustain. Built Environ. **94**, 150–176 (2015)
84. Alawadhi E.M, Using phase change materials in window shutter to reduce the solar heat gain. Energy Build. 421–429 (2012)
85. D. Buddhi. H.S. Mishra, A. Sharma, Thermal performance studies of a test cell having a PCM window in south direction. IEA, ECESIA Annex 17 (2003)
86. Mehling Harald, *Strategic Project "Innovative PCM-Technology"—Results and Future Perspectives, 8th Expert Meeting and Work Shop* (Kizkalesi, Turkey, 2004)
87. N. Soares, J.J. Costa, A. Samagaio, R. Vicente, Numerical evaluation of a phase change material—shutter using solar energy for winter nighttime indoor heating. J. Build. Phys. 367–394 (2014)
88. L. Shuhong, S. Gaofeng, Z. Kaikai, Z. Xiaosong, Experimental research on the dynamic thermal performance of a novel triple-pane building window filled with PCM. Sustain. Cities. Soc. 15–22 (2016)
89. C. Liu, Y. Zheng, D. Li, H. Qi, X. Liu, A model to determine thermal performance of a non-ventilated double glazing unit with PCM and experimental validation. Procedia Eng. 293–300 (2016)
90. G.M. Gomes, A.J. Santos, M.A. Rodrigues. Solar and visible optical properties of glazing systems with venetian blinds: numerical, experimental and blind control study. Build. Environ. 47–59 (2014)
91. Silva Tiago, Vicente Romeu, Amaral Cláudia, Figueiredo António, Thermal performance of a window shutter containing PCM: Numerical validation and experimental analysis. Appl. Energy **179**, 515–535 (2016)

92. Silva Tiago, Vicente Romeu, Rodrigues Fernanda, Samagaio António, Development of a window shutter with phase change materials: full scale outdoor experimental approach. Energy Build. **88**, 110–121 (2015)
93. Silva Tiago, Vicente Romeu, Soares Nelson, Ferreira Victor, Experimental testing and numerical modelling of masonry wall solution with PCM incorporation: a passive con-struction solution. Energy Build. **49**, 235–245 (2012)
94. Silva Tiago, Vicente Romeu, Amaral Cláudia, Samagaio António, Cardoso Claudino, Performance of a window shutter with phase change material under summer Mediterranean climate conditions. Appl. Therm. Eng. **84**, 246–256 (2015)
95. A. Castell, I. Martorell, M. Medrano, G. Pérez, L.F. Cabeza, Experimental study of using PCM in brick constructive solutions for passive cooling. Energy Build. **42**, 534–540 (2010)
96. A.V. Sá, M. Azenha, H. Sousa, A. Samagaio, Thermal enhancement of plastering mortars with phase change materials: experimental and numerical approach. Energy Build. **49**, 16–27 (2012)
97. H.J. Alqallaf, E.M. Alawadhi, Concrete roof with cylindrical holes containing PCM to reduce the heat gain 73–80 (2013)
98. L. Royon, L. Karim, A. Bontemps, Thermal energy storage and release of a new compo-nent with PCM for integration in floors for thermal management of buildings. Energy Build. **63**, 29–35 (2013)
99. X. Xu, Y. Zhang, K. Ling, H. Di, R. Yang, Modeling and simulation on thermal performance of shape-stabilized phase change material floor used in passive solar buildings. Energy Build. **37**, 1084–1091 (2005)
100. A.G. Entrop, H.J.H. Brouwers, A.H.M.E. Reinders, Experimental research on the use of micro-encapsulated phase change materials to store solar energy in concrete floors and to save energy in Dutch houses. Sol. Energy **85**, 1007–1020 (2011)
101. L. Royon, L. Karim, A. Bontemps, Thermal energy storage and release of a new component with PCM for integration in floors for thermal management of buildings. Energy Build. **63**, 29–35 (2013)
102. L. Royon, L. Karim, A. Bontemps, Optimization of PCM embedded in a floor panel developed for thermal management of the lightweight envelope of buildings. Energy Build. **82**, 385–390 (2014)
103. R. Ansuini, R. Larghetti, A. Giretti, M. Lemma, Radiant floors integrated with PCM for indoor temperature control. Energy Build. **43**, 3019–3026 (2011)
104. K.L. Huang, G.H. Feng, J.S. Zhang, Experimental and numerical study on phase change material floor in solar water heating system with a new design. Sol. Energy **105**, 126–138 (2014)
105. G.B. Zhou, J. He, Thermal performance of a radiant floor heating system with different heat storage materials and heating pipes. Appl. Energ. **138**, 648–660 (2015)
106. M. Zhao, T.T. Zhu, C.N. Wang, H. Chen, Y.W. Zhang, Numerical simulation on the thermal performance of hydraulic floor heating system with phase change materials. Appl. Therm. Eng. **93**, 900–907 (2016)
107. Y. Xia, X.S. Zhang, Experimental research on a double-layer radiant floor system with phase change material under heating mode. Appl. Therm. Eng. **96**, 600–606 (2016)
108. J.F. Belmonte, P. Eguía, A.E. Molina, J.A. Almendros-Ibáñez, Thermal simulation and system optimization of a chilled ceiling coupled with a floor containing a phase change material (PCM). Sustain. Cities Soc. **14**, 154–170 (2015)
109. A. Pasupathy, R. Velraj, Effect of double layer phase change material in building roof for year round thermal management. Energy Build. **40**, 193–203 (2008)
110. J. Kosny, K. Biswas, W. Miller, S. Kriner, Field thermal performance of naturally ventilated solar roof with PCM heat sink. Sol. Energy **86**, 2504–2514 (2012)
111. M. Koschenz, B. Lehmann, Development of a thermally activated ceiling panel with PCM for application in lightweight and retrofitted buildings. Energy Build. **36**, 567–578 (2004)
112. H. Weinläder, W. Körner, B. Strieder, A ventilated cooling ceiling with integrated latent heat storage—Monitoring results. 65–72 (2014)

113. J. Kosny, E. Kossecka, A. Brzezinski, A. Tleoubaev, D. Yarbrough, Dynamic thermal performance analysis of fiber insulations containing bio-based phase change materials (PCMs) 122–131 (2012)
114. Y. Lei, X. Zhang, G. Xu, Thermal performance of a solar storage packed bed using spherical capsules filled with PCM having different melting points. Energy Build. **68**, 639–646 (2014)
115. Xing Jin, Shuanglong Zhang, Xu Xiaodong, Xiaosong Zhang, Effects of PCM state on its phase change performance and the thermal performance of building walls. Build. Environ. **81**, 334–339 (2014)
116. M.A. Izquierdo-Barrientos, J.F. Belmonte, D. Rodríguez-Sánchez, A.E. Molina, J.A. Almendros- Ibáñez, A numerical study of external building walls containing phase change materials (PCM). Appl. Therm. Eng. **47**, 73–85 (2012)
117. Kuznik Frédéric, Virgone Joseph, Experimental assessment of a phase change material for wall building use. Appl. Energy **86**, 2038–2046 (2009)
118. G. Evola, L. Marletta, The effectiveness of PCM wallboards for the energy re-furbishment of lightweight buildings. Energy Procedia **62**, 13–21 (2014)
119. Jin Xing, Zhang Shuanglong, Effects of PCM state on its phase change performance and the thermal performance of building walls. Build. Environ. **81**, 334–339 (2014)
120. N. Sarier, E. Onder, Organic phase change materials and their textile applications: an overview. Thermochim 7–60 (2012)
121. K. Horikiri, Y. Yao, J. Yao, Numerical optimization of thermal comfort improvement for indoor environment with occupants and furniture 303–315 (2015)
122. X. Yang, P. Fazio, H. Ge, J. Rao, Evaluation of moisture buffering capacity of interior surface materials and furniture in a full-scale experimental investigation, 188–196 (2012)
123. L.H. Mortensen, C. Rode, R. Peuhkuri, Investigation of airflow patterns in a microclimate by particle image velocimetry (PIV), 1929–1938 (2008)
124. M. Corcione, L. Fontana, G. Moncada Lo Giudice, A parametric analysis on the effects of furnishings upon the performance of radiant floor-panel heating systems 59–68 (2000)
125. M.Z. Pomianowski, F. Khalegi,G. Domarks, J. Taminskas, K. Bandurski, K.K. Madsen, et al. Experimental investigation of the influence of obstacle in the room on passive night-time cooling using displacement ventilation 499–506 (2011)
126. L. Fontana, Thermal performance of radiant heating floors in furnished enclosed spaces 1547–1555 (2011)
127. J. Le Dréau, Energy flow and thermal comfort in buildings—comparison of radiant and air-based heating and cooling systems (2014)
128. K.A. Antonopoulos, E.P. Koronaki, Effect of indoor mass on the time constant and thermal delay of buildings 391–402 (2000)
129. J. Yam, Y. Li, Z. Zheng, Nonlinear coupling between thermal mass and natural ventilation in buildings, 1251–1264 (2003)
130. H. Wolisz, T.M. Kull, R. Streblow, D. Müller, *The Effect of Furniture and Floor Covering upon Dynamic Thermal Building Simulations* (2015)
131. P. Raftery, E. Lee, T. Webster, T. Hoyt, F. Bauman, Effects of furniture and contents on peak cooling load 445–457 (2014)
132. Q. Nguyen, T. Ngo, P. Mendis, A review on fire protection for phase change materials in building applications, in *From Materials to Structures: Advancement Through Innovation*, ed by Samali, Attard, Song (Taylor & Francis Group, 2013)

Chapter 4
Conclusions

It was carried out a review of thermal energy storage using phase change materials with focus on the building application. The information gathered is divided into the different application of PCM. From the research, it can be concluded that PCM application for passive solutions in construction materials has been studied for a couple of decades and by many academics.

There are several materials that can be used as phase change materials. A common way to distinguish them is by dividing into organic, inorganic and eutectic PCMs. These categories are divided based on the various components of the PCMs. Paraffin and binary organic acids are the main phase change materials used in envelopes.

Flammability risk is still an issue so all products with PCM shall be tested with respect to fire retardation and accomplish necessary fire codes and standards. Suitable thermal properties of PCMs and their composites, such as thermal conductivity, thermal diffusivity, specific volumetric heat and specific heat capacity, have to be determined as a function of temperature to properly determine dynamic performance and potential for the entire energy storage system. Regarding measurements of thermal properties of PCMs and their composites, some experimental methods are presented. For specific heat capacity measurements, DSC isothermal step mode, DTA and T-history methods can be recommended. On the contrary, results from DSC dynamic mode might be dependent on sample size and temperature ramp.

Heat stored in the PCM during high-temperature periods (days) should be discharged during low-temperature periods (nights) to be able to perform on the consecutive day, so phase change materials can decrease energy consumption, shift the peak loads of cooling energy demand, decrease temperature fluctuations providing a thermally comfortable environment and reduce the electricity consumption. The thermal inertia, which can be defined as "time lag" and "decrement factor", is one of the important parameters to estimate the thermal performance of the buildings.

© The Author(s), under exclusive license to Springer Nature Switzerland AG 2019
J. M. Delgado et al., *Thermal Energy Storage with Phase Change Materials*,
SpringerBriefs in Applied Sciences and Technology,
https://doi.org/10.1007/978-3-319-97499-6_4

When studying the potential of PCM products, boundary condition (temperature fluctuation, heat transfer on the surface) and heating loads in experimental set-ups and in simulations have to represent realistic condition; otherwise, their performance will be either over or underestimated. When used in buildings, PCMs can be incorporated into other building materials. This topic has attracted a lot of interest as it will allow buildings to be built in a similar way that they are built today, but with materials that have an increased thermal energy storage capacity. In the reviewed literature, it was recognizable that wall systems integrating PCMs have received the most attention.

As most PCMs designed for building applications go through a liquid phase, encapsulation is needed to avoid complications such as leaking of PCM to the surface and spreading of low viscous liquids throughout the material. Therefore, methods such as direct incorporation and immersion of PCMs in building materials are not well suited for long-term applications.

Nevertheless few detailed studies on the overall effect of PCMs in real-life constructions, commercial PCM products have already been used in several projects.

Gypsum materials can be combined by up to 45% by weight of PCM when strengthening the structure with some additives and up to 60% by weight in the wallboard composites. On the other hand, in the concrete materials only up to approximately 6% (by weight) of PCM could have been implemented.

The PCM in the glazed envelopes, only few researches can be found and much less work has been documented comparing to the opaque constructions with PCM. Based on the reviewed studies, it can be concluded that PCM in glazing, shading and shutters might be an interesting addition to the building envelope in order to minimize solar heating loads. Application of PCM in the glazed surfaces shall be carefully designed, since on one hand, it can help reduce solar thermal loads during the hot season, but on the other hand, it can decrease thermal resistance of windows and by that increase heat losses during the cold season. Therefore, potential of PCM in the glazed envelopes should be carefully studied with respect to the climate condition.

PCM-ceramic tiles present remarkable potential for improving the thermal comfort inside buildings due to their ability to reduce indoor space temperature fluctuations.

The accurate potential to increase dynamic heat storage capacity of concretes by incorporation of PCM is doubtful. Firstly, the thermal mass increase is not as high as expected, and secondly, thermal conductivity decreases significantly due to addition of PCM to concrete. Only, 5–6% by weight of PCM corresponds to approximately 12–15% by volume of concrete, which means that the share of PCM in concrete is rather high and as a result, the price of the composite would be high due to rather high price of PCM.

Furnishing elements offer a large surface area exposed to the indoor environment, which makes them a good candidate for PCM integration. It can be an interesting solution for the implementation of passive LHTES systems without construction work and thereby improving thermal inertia and energy flexibility of light buildings. However, the integration of PCM in furniture raises new issues concerning their defiance with fire regulation, recycling process and total life cycle analysis.

It can be concluded that potential of latent heat application should always be compared to realistic and relevant reference systems/constructions, heat storage potential of PCM technologies should always be analysed taking into account thermal properties of PCMs/PCM composites but also heat transfer condition on the surface.

4.1 Further Suggestions

Further investigations still need to be carried out on the incorporation methods for PCMs to be embedded in existing building structures, long-term stability and any other problems which may affect the safety, reliability and practicability of the thermal energy storage used in buildings.

In published studies, most of the information concerning high-temperature PCMs lacks completeness. Consequently, it is necessary for researchers to attach importance to completing and providing various aspects of information when actively conducting experiments to develop new materials. The most important items of information are thermal conductivity in the operating temperature range, specific heat in different phase states, environmental impacts, costs and other factors, so as to be able to measure each material's application value more comprehensively and objectively.

There is no standard method (such as British Standards or EU standards) developed to test for PCMs, making it difficult for comparison to be made to assess the suitability of PCMs to particular applications. A unified platform such as British Standards, EU standards needs to be developed to ensure same or similar procedure and analysis (performance curves) to allow comparison and knowledge gained from one test to be applied to another.

In PCM research community, there is a lack of one TA method convenient to measure larger PCM samples. T-history is the suitable candidate. To move towards a commercial and available one is important to get in consensus among all researchers to suggest a common instrumental setup, data analysis and presentation of final results.

Printed in the United States
By Bookmasters